PERSONAL SUCCESS STRATEGIES

Developing Your Potential!

MEL HENSEY

A HANDBOOK

Published by

American Society of Civil Engineers
1801 Alexander Bell Drive
Reston, Virginia 20191-4400

Abstract: This handbook for managers and leaders presents strategies used by individuals who have experienced professional success and personal fulfillment. In this volume, the author focuses on the individual and shares strategies that have been successful for him as well as for his clients. Strategies offered are based on recent research and focus on attitudinal or habitual behaviors and are designed to assist individuals in making changes and developing themselves to be more effective and successful.

Library of Congress Cataloging-in-Publication Data

Hensey, Mel.
Personal success strategies : developing your potential! : a handbook / Mel Hensey.
p. cm.
Includes bibliographical references (p.).
ISBN 0-7844-0446-1
1. Civil engineers—Vocational guidance. 2. Success in business. 3. Personal relationships. I. Title.
TA157.H455 1999
624′.068—dc21 99-41071
CIP

The material presented in this publication has been prepared in accordance with generally recognized engineering principles and practices, and is for general information only. This information should not be used without first securing competent advice with respect to its suitability for any general or specific application.

The contents of this publication are not intended to be and should not be construed to be a standard of the American Society of Civil Engineers (ASCE) and are not intended for use as a reference in purchase specifications, contracts, regulations, statutes, or any other legal document.

No reference made in this publication to any specific method, product, process or service constitutes or implies an endorsement, recommendation, or warranty thereof by ASCE.

ASCE makes no representation or warranty of any kind, whether express or implied, concerning the accuracy, completeness, suitability, or utility of any information, apparatus, product, or process discussed in this publication, and assumes no liability therefor.

Anyone utilizing this information assumes all liability arising from such use, including but not limited to infringement of any patent or patents.

Library of Congress Catalog Card No: 99-41071
ISBN 0-7844-0446-1
Manufactured in the United States of America.

FOCUS OF THE BOOK

Individuals are (individually) unique! We each have different "body suits," roles, values, needs, beliefs, goals and journeys. Our journeys may be sad or joyful, long or short, noteworthy or anonymous.

> **The aim of this handbook is to draw on the wisdom of many resources and offer some strategies for learning and for increasing individual effectiveness and career satisfaction, no matter what your purpose or goals in life may be.**
>
> - **"Success" in this book is intended to be defined primarily by the reader's own goals, hopes, criteria and environment.**
> - **It is organized (I hope) in such a way that the reader can find a needed topic quickly.**
> - **Readers who want more information on a topic are offered additional resources in many cases.**

Brian Tracy (***The Psychology of Achievement*** [tapes]; Nightingale-Conant, Niles, IL; 1994) ruefully observes that most of us come into life with a **complex operating system** called a body/mind and **no operator's manual!** He also notes that many of us never really learn how to use it well until we are in middle age.

> **We can do little to change the past, but we have more power than we realize to mitigate the past and to shape our future, as individuals and collectively.**

NOTES AND CAUTIONS

> **"Time is the greatest teacher. Unfortunately, it kills all its students."**
>
> **... Jim Holt, *Management Review***

Some of the **strategies** that "successful" people have used to create their own professional success and personal fulfillment have long been of interest to me and may be useful to you.

My objective is to add to **your repertoire and toolbox.** Individuals with more options have a much greater chance for "success" ... however they personally define success!

- **These ideas and strategies may or may not "fit" for you, or work for you. You must consider and decide what, if anything, you will sample, explore or try for yourself!**
- **They are not intended as a substitute for professional assistance you may need, such as personal counseling, career counseling, health counseling, crisis counseling, legal or financial guidance or any other counseling need.**
- **In fact, one strategy that successful people often use, is to seek professional guidance or counseling or coaching in order to develop a particular skill set, or to deal with a particular need or situation!**

... Mel Hensey, 1999

WHY I WROTE THIS BOOK

Collective Excellence came into being because ASCE asked me to do it, and because **teams** are so important yet so underutilized in today's everyday work environments.

ASCE also asked me to write ***Continuous Excellence***. I believed it was important because of the many misguided, expensive and often harmful organizational "fixes" being promoted across the land.

We live in challenging times! And so, when ASCE asked me to do another book, I thought perhaps the **Individual** needed to be the focus of this one. I took this perspective: What do I understand **now** that would have been very helpful to have known sooner?

So, **the topics** selected for this book come from my observations of what would have been helpful to me in my own life experiences and those of our consulting clients, our friends and family. And, I hope it will be truly useful to you.

The 20 chapters address some of the more common everyday puzzles and problems of life and work, about which I may have ideas to offer. There are many areas where I'd like to comment but cannot!

If you have suggestions or feedback, I'd be happy to have them come to me at ...

- 8220 Rivers Edge Circle, Maineville, Ohio 45039
- E-mail: mhensey@aol.com

Thanks and may your life and work go well!

Mel Hensey, PE, F.ASCE
Management Consulting Engineer, Maineville, Ohio, 1999

ACKNOWLEDGMENTS

Many authors, teachers, colleagues, clients and friends have contributed to almost every section of this handbook. Where feasible, I've acknowledged these contributions.

Yet, I also know that many other people, forgotten and/or too numerous to list here, have influenced my life and development, and the learnings shared here.

Specifically, I want to express my gratitude to ...

- **The Lord of Life**, for the opportunity to live, to learn, to enjoy this earth and my fellow pilgrims, and to contribute.

- **ASCE**, for asking and for pursuing this work in spite of my procrastination.

- **Carol Hensey**, my partner and spouse, for patient support in so many ways over many years.

- **John Parmater**, for influencing my consulting work and my writing style.

- **Many of our clients**, whose questions and situations have stimulated my consulting and my writing.

- **Ann Somboretz**, for talented word processing, formatting, graphics and editing.

- **The authors mentioned** in particular chapters, for making the effort to share their insights and wisdom.

TABLE OF CONTENTS

It's difficult to slice any holistic area like "personal development" into "chunks" and use linear thinking. Each chunk or topic relates to many others. So I've arranged these chapters in the style of a **handbook** to help the reader more easily locate topics of interest:

FOREWORD

Many sections of this reference book attempt to assist individuals in making changes and developing themselves to be more effective and/or successful, in whatever ways they want.

I've noticed that changing **some** things—even a little—can be difficult and frustrating. Based on my own learnings as well as those of friends, I've examined what it is that is challenging to change.

Here's a "menu" of relative difficulty for makingchanges ...

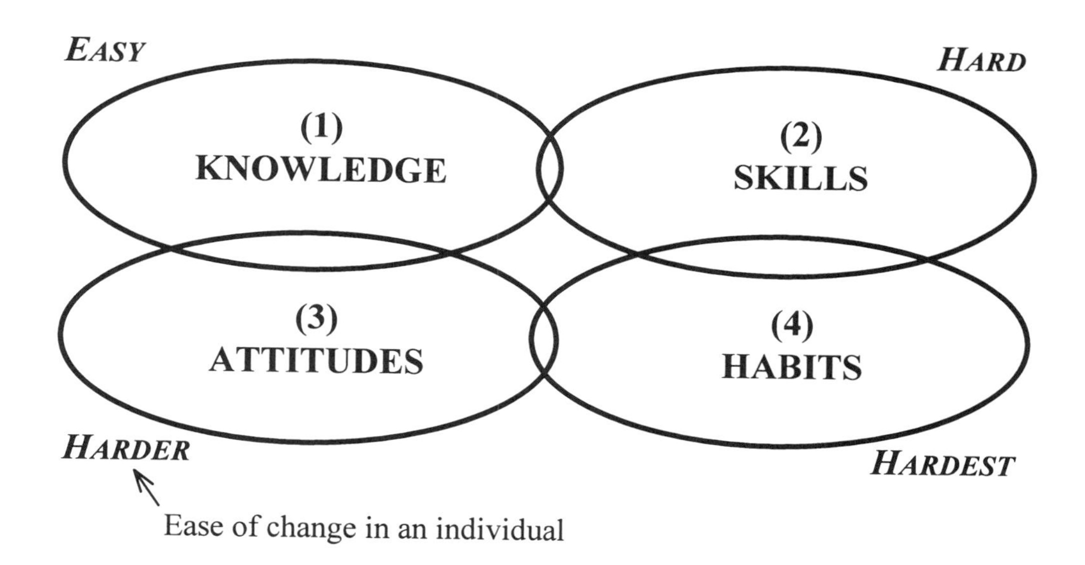

Recent research suggests that a very high percentage of our daily behaviors are programmed by the **unconscious** mind. They are quite challenging to change even if we want to or ought to!

Many of the ideas offered in the following chapters are designed to assist with insight or cognitive processes. And, some ideas may also focus on attitudinal or habitual behaviors the reader wants to change for him or herself.

Chapter 1

MENTAL MODELS AND MAPS

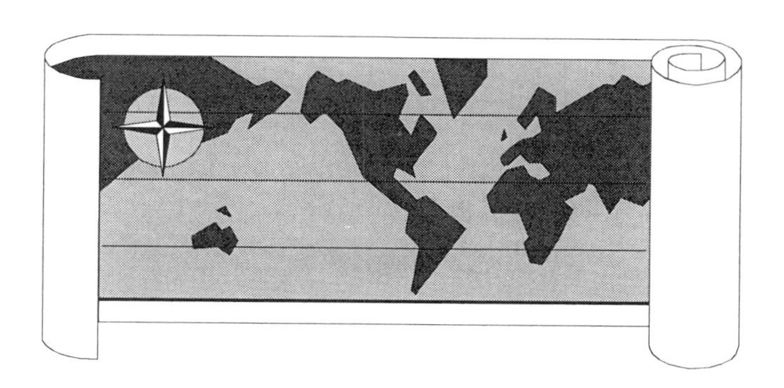

We owe a great debt to the people who've helped us begin to realize that we live in a world that is ... to a very great extent ... created in our own unique **individual** and **group** imaginations. And as we come to more fully understand this and its implications, we better understand others, the mysteries of life and some of the other notions in this book.

> **Each of us has a unique mental "model" or "map" of how the particular parts of the world behave or ought to behave. These particular parts might include: the realm of business, the realm of religion, the realms of child rearing or marriage, the realms of government and politics, even the realm of health.**

Alfred Korzybski has perhaps given us the best terminology to describe these beliefs and thought processes in a helpful, clear way in his now classic book ***Science and Sanity*** (originally published in 1933, republished in 1995 by the Institute of General Semantics, Englewood, NJ), where he notes that "words are not things, maps are not the actual territory."

His work is credited by the early developers of Neuro-Linguistic Programming (NLP) and in it he coined the term "neuro-linguistic" and used the map-territory distinction (***Anchor Point Newsletter***, May 1992).

Korzybski notes that we humans are **"semantic creatures"** due to the very way our nervous system operates; it abstracts, generalizes, summarizes and symbolizes the world around us. **And**, we may fail to distinguish between symbol and reality.

His work leads us to some practical **guidelines** for daily living, working, communicating, learning, deciding and so on:

- Each one of us has our own unique **map** of "reality."
- Our maps are **not** the territory (reality), only maps of it.
- Each of our maps are unique and **differ** from the maps of other people.
- People respond to their **maps** of reality, not to reality itself.
- We tend to see the errors in **others'** maps more easily than we see the errors in our **own** maps. We tend to overvalue our own maps vs. others' maps.
- People always make the best choices available to them at the time, given the **limits** of their maps.
- To be understood by another person, we must speak to **their** map of the world.

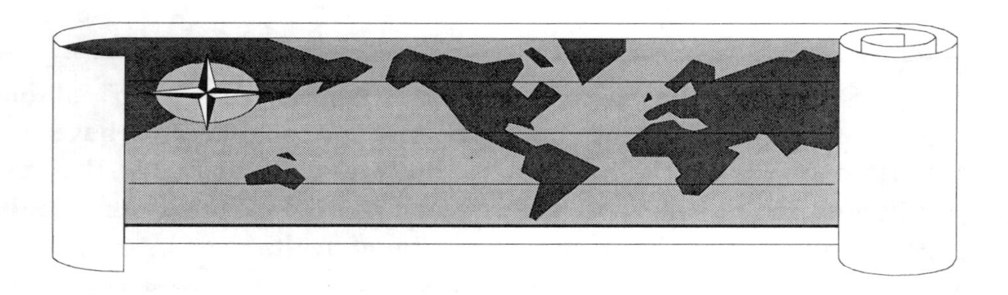

Also, over the centuries, many thoughtful people have told us that …

- We are limited by our beliefs and opinions, and by what Mark Twain calls **"what we 'know' that just isn't so!"**
- We can achieve what we hope for and dream of if we change our belief systems to allow ourselves to "see" the possibilities …

  … or reversing the doubters cliché "I'll believe it when I see it" to … **"I'll see it when I believe it."**

Possibility Thinking

Dawna Markova (***The Art of the Possible***, Conari Press, Berkeley, CA, 1991) has pulled together some helpful thoughts of others. These observations focus on our limiting maps and beliefs and, more importantly, on the **possibilities** beyond them. Here are some of my favorites …

- "Alert, waking consciousness, your ordinary state, your cultural trance, is when we all dream the same dream, more or less, and call it reality."

  … Robert Masters and Jean Houston
  Mind Games (Quest Books, 1998)

- "The greatest enemy of any one of our truths may be the rest of our truths."

  … William James

- "Education consists mainly in what we have unlearned."

  … Mark Twain

- "The voyage of discovery lies not in finding new landscapes, but in having new eyes."

  … Marcel Proust

- "Whatever authority I may have rests solely on knowing how little I know."

  … Socrates

- "Tell me what you pay attention to and I will tell you who you are."

  … Jose Ortega y Gasset

- "In order to understand what another person is saying, you must assume it is true, and try to imagine what it is true of."

  … George Miller, cognitive scientist

- "The test of a first-rate intelligence is the ability to hold two opposed ideas in the mind at the same time, and still retain the ability to function."

  … F. Scott Fitzgerald

- "The most beautiful thing we can experience is the mysterious. It is the source of all true art and science."

  … Albert Einstein

Key Points

I'm presenting the human phenomenon of mental models and maps in Chapter 1 because …

- **Your** map will affect what you'll get out of all the following chapters.
- One of my goals is to **broaden and enrich** your mental map(s) of "reality."
- If you are more attuned to others' maps as **maps** (including this book), you are more likely to be successful.

Exploring Your Map(s)

As noted on page 1, each individual will actually have many maps. Take a moment now and check **your map** of "success."

What does "success" mean to you personally? That is, how do you define it? What is required to be successful? Who determines if you are successful? And so on …

Chapter 2

STAGES OF HUMAN DEVELOPMENT

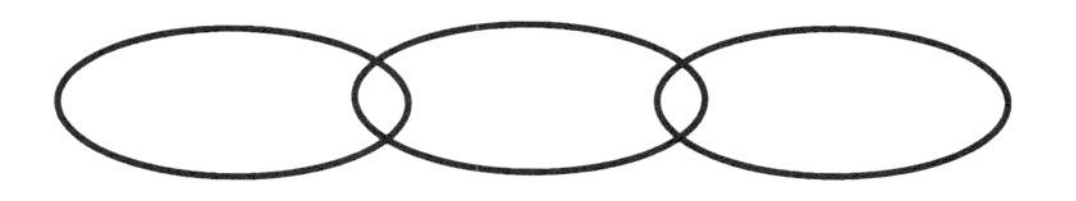

Why is this topic important enough to be one of the first sections in a reference book on personal excellence?

Because all of us generally go through several rather **predictable stages** of life that are patterned by powerful unseen forces, including

- genetic coding, and
- cultural conditioning.

This knowledge may also enable us to make **wiser choices** in living life, particularly when some of these stages may be frustrating, confusing or feel like crises in our lives.

On a personal level, at age sixty-four, I can now say it would have been a great comfort to know that some of the difficulties and hassles, foolishness and madness, confusion and crises were things we **all** experience to one degree or another, one way or another.

> The difficulties of job and career change, health problems, marital challenges, and the stress of active and wonderful children, were pretty "normal" but I didn't know it! And, many friends didn't either.

Stages of Adult Development

In 1974, Gail Sheehy's book ***Passages*** (Bantam Books, NY) became popular and well read. Her descriptions of the inevitable(!) stages of human development, including certain predictable crises, were fascinating to many.

Before that, most of us thought we only grew through "stages" as children and youth … and from then on we simply grew **older** as adults!

These adult stages have fewer visible "markers" such as the changes in size, shape, hair, voice and puberty of youth. But there are other external markers (like graduation, careers, marriage, settling, uprooting, etc.) which reflect the changes **within**.

Each adult stage presents new challenges of development that require **letting go** and also trying **new approaches** because the old ones no longer work as well.

Here are the stages Sheehy identified in her colorful descriptions:

- **Pulling Up Roots** (unplugging from home and family)
- **The Trying Twenties** (learning to cope in an adult world)
- **Catch-30** (feeling pinched or constrained with one's life structure around the age of 30)
- **Deadline Decade** (focus on work, family and achievement in the mid to late 30s)
- **Renewal or Resignation** (mid 40s reflection and look at the "rest" of one's life)

More on Adult Stages

Daniel J. Levinson did biographical research on men's lives culminating in his book ***The Seasons of a Man's Life*** (Ballantine Books, NY, 1978), particularly focusing up to age 45. He identified with Carl Jung's idea of a mid-life transition from the first to the second half of life around 40.

Perhaps the greatest student of human behavior in modern times, Carl Jung, noted that, until the late 30s, our existence is so focused on coping with the demands of life that some parts of our personalities are necessarily neglected.

Jung identified four psychological functions … **thought**, **feeling**, **intuition** and **sensation** … not equally developed. He believed they can and should be developed in the "second half" of life.

In the '80s, Dan and Judy Levinson studied women's lives, and then in 1996, ***The Seasons of a Woman's Life*** was published (Ballantine Books, NY). In it they "made the surprising discovery that women and men go through the same sequence of periods at the same ages." Well!

The authors also noted wide variations between and within the genders, as well as how individuals "traverse" these stages of life.

The work of both Levinson and Sheehy indicates that we must wrestle well with the internal developmental tasks brought by each "stage" or the crises may return at a later stage with renewed importance and impacts.

Four Stages of Professional Careers

In the world of our daily **work**, Dalton, Thompson and Price provide us with important findings on the stages of professional development (see *Table 2.1*). If they are successful, most professionals go through these four stages, as reported in ***Organizational Dynamics*** (an AMACOM Journal, Summer, 1977):

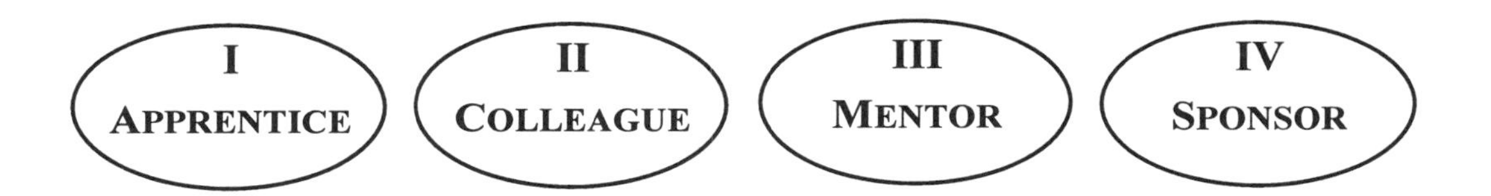

Table 2.1: Stages of Professional Development

APPRENTICE STAGE I	COLLEAGUE STAGE II
○ Works under the supervision and direction of a more senior professional in the field. ○ Is given assignments which are a portion of larger projects or activities. ○ Is overseen by senior professionals. ○ Lacks experience and status in organization. ○ Is expected to exercise "directed" creativity and initiative.	○ Goes into depth in one primary technical area. ○ Assumes responsibility for a definable portion of the project, process, or enterprise. ○ Works independently and produces significant results. ○ Relies less on supervisor or mentor for answers; develops own resources to solve problems. ○ Develops credibility and a reputation.

Table 2.1: Stages of Professional Development *... continued ...*

<table>
<tr><th>MENTOR
STAGE III</th><th>SPONSOR
STAGE IV</th></tr>
<tr><td>○ Makes significant technical contributions while working in several areas.
○ Offers greater breadth of technical skills and application of those skills.
○ Stimulates others through ideas and information.
○ Is involved in developing people in one or more ways:
• acts as an idea person for a small group,
• serves as mentor to younger professionals,
• assumes supervisory or leadership position.
○ Works outside in relationships with client organizations, developing new business, etc.</td><td>○ Influence gained through:
• past ability to assess environmental trends,
• ability to deal with the outside effectively,
• ability to affect others inside the organization.
○ Influences future direction of his/her organization by means of:
• original ideas, leading the organization into new areas of work;
• organizational leadership and policy formation;
• integrating the work of others.
○ Engages individuals and groups both inside and outside the organization.
○ Sponsors development of promising people for future key roles.</td></tr>
</table>

While Dalton, et. al focused their research primarily on engineers and scientists, their conclusions seem to serve very well for many professionals. My colleagues and I have noted similar stages of professional development for the fields of

- medicine and health care
- information technology
- marketing and sales
- all areas of education

- government and public service
- manufacturing professionals
- banking and finance.

We believe work is such an important part of life for so many years, that these **stages** of professional development are vital to making the most of life and, certainly, careers.

As you think about your own life and career—past, present and future—and your work activities:

- **What "life stage" are you (mostly) in now?**

  ___

  ___

  ___

- **What stage of professional development are you (mostly) in now?**

  ___

  ___

  ___

- **What kinds of work are you doing that may lead into the next stage?**

  ___

  ___

  ___

  ___

  ___

  ___

  ___

  ___

  ___

  ___

  ___

  ___

- **Other thoughts about life stage(s) or career stage(s) for yourself or those you are close to …**

Chapter 3

YOUR LIFE GOAL, AIM OR PURPOSE

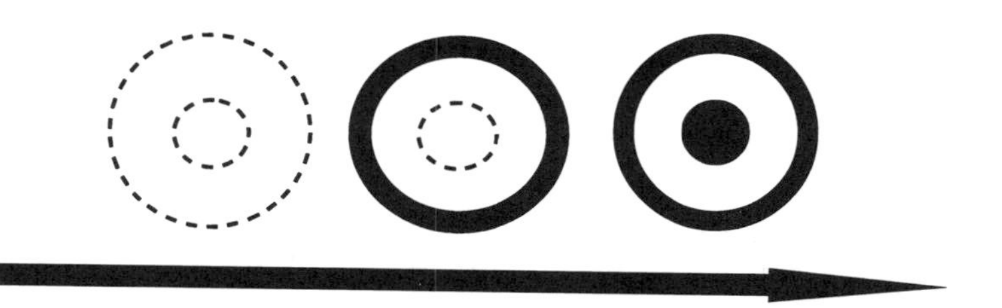

> *Our very business in life is not to get ahead of others, but to get ahead of ourselves.*
>
> ... Thomas Monson

When most people have the opportunity to ponder deeply about what they really want, they almost always want something a bit bigger or a bit different than they have or than they are **now**.

For institutions (companies, colleges, churches) the popular term for that "want" is their **vision**, or their collective desired future state.

For individuals, the terms **life goal, aim, purpose or mission** seem to work pretty well. This life purpose may seem elusive, rather like a vague longing or dream, almost too deep for words to describe.

When I'm working with **a group**—a board, a planning team, an executive committee or such—the group's interaction often provides the stimuli needed to spark these hopes and dreams and draw them out where they can be discussed to form a vision or a plan.

When I'm working with **an individual**, I share some helpful questions that I've collected over the years from many sources. Following are some questions that have helped **individuals** evoke some of their hopes and dreams.

Great Questions

⊙ Inspired by AMA colleague Dr. Ann O'Roark's "Outback Walk-about" …

- My proudest **achievement** has been: ______________________________

- At this stage in my life, the most **important** thing to me is: ______________________________

- If I had my "druthers," my **dream** would be to: ______________________________

- I would most like to be **remembered** for: ______________________________

⊙ Inspired by Lyman Coleman's "Serendipity" workshops, these questions are about heroes and heroines (living or dead, real or fictional) …

- Who were my **childhood** heroes? ______________________________

- Who are some of my heroes **today**? ______________________________

- Who are some of the people who helped **shape** my life? ______________________________

- In my profession or business, whom do I regard with great **respect**? _____

- When I'm doing my very best, whom do I wish **knew** about it? _________

⊙ More good questions that sometimes draw forth deep hopes …

- What were/are my **parents'** expectations of me? ____________________

- Who are my favorite **characters** from TV, films and books? __________

- Who am I? (Describe **yourself** without resorting to roles such as mother or nurse.) ____________________

- Who do I **pretend** to be, when I'm pretending to be different than I am?

- Who do I **want** to be? ____________________

Skills and Activities

Many of us engage in activities and/or use skills that are **deeply satisfying**. If we're lucky, we do them at our work (our "day job" as the saying goes). Or, maybe we do them in our hobbies and outside interests. Or perhaps both.

Think about some of your most satisfying activities whether at "work" or as avocations:

What are the most satisfying **activities** you do, or **skills** you use?	What is there about doing it that is so **satisfying**?
________________________	________________________
________________________	________________________
________________________	________________________
________________________	________________________
________________________	________________________
________________________	________________________
________________________	________________________
________________________	________________________
________________________	________________________
________________________	________________________

Gaining Clarity

From questions like these, you can probably tease out some important hopes you have for your life. It won't happen quickly or easily, but it is often enjoyable to work on! As you think about these questions, your life purpose may become progressively clearer to you over time, like so:

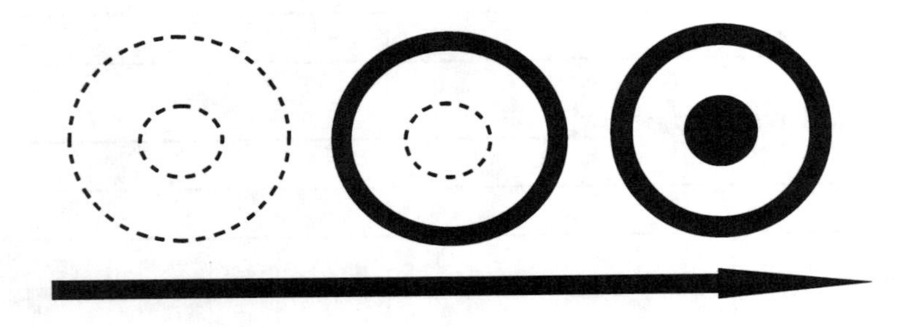

If you are among that very small percentage of people who think about or who know what you want, this chapter will have been reassuring for you.

Betty Eadie learned a great deal from her near-death experiences, which she shared in her book ***Embraced by the Light*** (Gold Leaf Press, Placerville, CA, 1992, available in paperback).

Other writers and researchers on the experiences of near or brief death have had similar findings which I find fascinating and reassuring. Here are some common themes:

- ⊙ Parts of **everyone's** purpose seem to include ...
  - gaining true wisdom, and
  - contributions to others.
- ⊙ Other parts of our individual purposes are **unique to us**.
- ⊙ Increasing **clarity** about our purpose may come at any stage of our lives. For example, Mother Teresa's purpose became clear to her quite late in life.

Your Personal Mission

Stephen Covey, author of ***The Seven Habits of Highly Effective People*** (Simon & Schuster, NY, 1989), has suggested that almost everyone will benefit by trying to write their **personal mission statement**, purpose or aim. I can only agree! Why not give it a shot? If you've tried or done it before, it may be clearer now. If you haven't, consider this a "first draft."

≺ Mission (First Draft) ≻

____

____

____

____

____

____

____

The paths to the future are not found but made, and the act of making them changes both the maker and the destination.

... Warren Bennis

Chapter 4

INTERESTS, STYLES, VALUES AND "IQ"

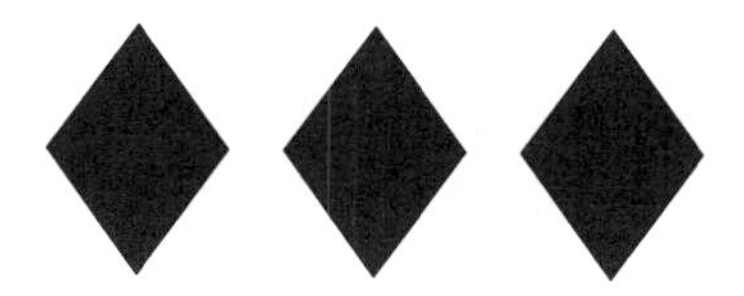

People are as different from one another as fingerprints or snowflakes! Even identical twins have significant differences. Carl Jung, pioneer in studying human behavior, believed that we each become even more unique as we grow through life. He called this process "individuation."

We must each deal with the demands of our culture and our institutions (such as schools) as well as the expectations of parents. And all the while, trying to first **discover** who we are, then to **develop** ourselves into

- someone **we want to be**
- a person who's acceptable to our **family**
- an individual that **society** finds "OK," preferably useful!

Good grief! No wonder it's often painful and somewhat unsuccessful. This is a real challenge for most of us.

Developing Interests, Styles and Values

In the process of individuation, we each develop our own interests, styles and values. Many activities and processes help us produce those results, and sometimes the processes are "invisible" to us (outside of our conscious awareness).

Here are some of the important processes of our own development that I've become aware of:

- **Observing our own family up close.** (Dad always cleared his throat before saying something difficult for him to say.)
- **Facing consequences of our initiatives.** (I well remember my mother's action on hearing my first swear word!)

- **Becoming aware of our own tastes.** (How could my dad actually like fried green tomatoes?)
- **Learning from others involved in our daily living.** (I loved to watch Tony set odd-shaped patio stones in mortar with skill, art and speed.)
- **Being reinforced for doing something we enjoy.** (My high school biology teacher said he liked my sketches a lot, my first "win" in school work.)
- **Sticking to a challenging task or skill set.** (After hours of confusion and frustration, I still recall the thrill of "seeing," finally, the trick of calculus.)
- **Being mentored, coached, guided or helped by another.** (Never too busy to answer a question, my helpful colleague Monte never made me feel stupid.)

Interests and Career "Fit"

I wish I could recall the name of the human resources manager who told me he believed that each person is **uniquely fitted** to do a particular job better than anyone else. There's a lot of evidence to support his belief.

People who change jobs, firms or careers, whether prompted by layoffs or their own initiative, often seem to be far happier as a result. (I said happier, not necessarily richer.) Sometimes it takes until mid-career to find that excellent niche where **interest**, **opportunity** and **usefulness** come together.

Speaker Tony Robbins likes to use the example of Colonel Sanders, who was well into retirement when he finally found a way to market his now famous fried chicken recipe.

There's a clear correlation with success for people who can satisfy at least some of their interests on-the-job as well as off. Human resources professionals use the term "motivated skills" for those abilities we **like** to use.

> In the long run, it benefits everyone for each of us to find a way to use our motivated skills in a way that helps our planet and/or its people.

Values and Style

Style and values actually go together, very much like compressed air in a pneumatic tire. The only values that mean very much are those that get acted upon ... **values in action**. And so, much of our (external) style is a function of our (internal) values.

Every job, career and role in life has a corresponding set of values that make for more or less success. And some values may even be prerequisites for a successful situation in life. These we might call **core values**, those values that are critical to success.

So ... what's a **value**? A prized belief. A guide to acceptable behavior. A norm, custom or implied rule. Here are a few samples of values:

- honesty and integrity
- sensitivity toward others
- punctuality
- perseverance

Values can often be in conflict! This is the case with the first two samples above; sometimes you cannot be both honest and sensitive.

Sometimes the only way we learn about values is by breaking one, often accidentally. Then our family or friends or work colleagues will usually let us know we are "out of line."

It **is important** to know the values that go with a particular profession or job or organization even if some of them don't make sense. Then we can at least know where the pitfalls are. And later, we may decide to try and change them to more appropriate values.

What's New in "IQ"?

Well, a lot! ***US News and World Report*** (November 23, 1987), among others, reports Harvard and Tufts research based on Howard Gardner's practical theory of **multiple realms of intelligence**.

"Getting Smart About IQ" is their report in which Gardner identifies seven ways to be bright, as follows:

- **Linguistic.** Language skills include a sensitivity to the subtle shades of the meanings of words.
- **Logical-Mathematical.** Both critics and supporters acknowledge that IQ tests measure this ability well.
- **Musical.** Like language, music is an expressive medium—and this talent flourishes in prodigies.
- **Spatial.** Sculptors and painters are able to accurately perceive, manipulate and recreate forms.
- **Bodily-Kinesthetic.** At the core of this kind of intelligence are body control and skilled handling of objects.

- **Interpersonal.** Skill in reading the moods and intentions of others is displayed by politicians, among others.
- **Intrapersonal.** The key is understanding one's own feelings—and using that insight to guide behavior.

Current IQ tests actually sample only a tiny aspect of human abilities. Gardner says: "While IQ tests may be good predictors of academic success, they are **poor** predictors of job success." Albert Einstein, who struggled with schooling, certainly wouldn't be surprised!

In the early '70s, while serving as a human resources manager for a large corporate engineering division, I plotted individuals' IQ type hiring test scores against their on-the-job work evaluations for a three year period—zero correlation! We stopped using those tests.

This also explains why managers who look for high IQs in their new hires, rather than a proven work track record, are often puzzled and disappointed.

So, we need to develop **all** of our skills and abilities—for work or fun—and forget about IQ and IQ type tests.

In the next several chapters, I'll offer more thoughts about skills and abilities that are important to career success as well as personal fulfillment.

But, before you turn to the next section, use this opportunity to jot down your **"motivated skills and abilities"**—that is, those capabilities that you **enjoy** using.

≺ Motivated Skills and Abilities ≻

Chapter 5

CORE COMPETENCIES FOR LIVING

I've often listened to a common dilemma faced by managers and leaders in our client organizations. It goes like this: A particular staff member is quite excellent, technically speaking. The employee, however, lacks some basic social skill such as ... appropriate language, tact, punctuality, or dressing appropriately.

The dilemma for the manager is that these basic and important social skills are the most **difficult** to talk about and request improvement in!

There are several lenses through which we can look at these "core competencies" or basic skills for living and working in a complex world, a world that is often crowded with other humans on whom we depend for various things.

Getting Along

The first lens would be one through which family members, work team colleagues or clients might see us. And here are some core competencies they'd be likely to expect from us:

- **Friendliness:** the absence of an "attitude."
- **Trustworthiness:** basic personal integrity and honesty.
- **Rapport:** able to put people at ease and comfortable to be with.
- **Basic humility:** humanness, a lack of arrogance, a sense of humor about oneself.
- **Acceptance of others:** open to human diversity in all its various aspects.
- **Assertive communication:** neither passive and whiny, nor aggressive and pushy.

Shoulder to Shoulder

This lens might be important again to family members and work colleagues, but certainly also to one's managers or business partners:

- **Perseverance:** doing whatever it takes to get the job done or to solve a problem.
- **Composure:** relative calmness and resourcefulness under difficult conditions.
- **Self-awareness:** understanding oneself, one's own biases, frailties and impacts on others.
- **Authenticity:** basic honesty about "who you are;" lack of pretense.
- **Relationships:** ability to build new relationships and mend frayed relationships when necessary.

Success Formula

Tony Robbins is a popular teacher and speaker on learning to use your "bio-computer" (mind) for your own success. Tony has identified a formula he believes all successful people instinctively use to achieve success in their chosen fields. Here's a brief summary of his findings:

- **Goal ...** have a goal (or several goals) and a plan.
- **Action ...** actually take action, take a first step toward the goal(s). (Few people do.)
- **Acuity ...** sensory acuity is required to note how well or poorly that action succeeds, and what "correction" is needed.
- **New Action ...** based on what we learned in step three, incorporate correction(s) into the next action.
- **Acuity ... again! ...** because perseverance is likely to be necessary for real success.

These steps can be enjoyable if ...

[1] We approach the "Acuity" step as a **detective** might, using all our senses carefully to determine what it was that didn't work well.

Approach it with intense curiosity instead of annoyance!

[2] We approach the "New Action" step as an **inventor** might, using all our creativity to craft the new action in a more effective way!

Getting Feedback

A new professional service emerging these days is called **professional coaching**. Sports or drama coaches have been around a long time of course, but these new coaches focus on everyday **working** skills such as sales, leadership, strategy, creativity or teamwork.

The thing these professional coaches offer that's so valuable is simply **objective, specific, behavioral feedback** with tips on how to improve in small ways. Here's an example of how it can work:

> I played my first golf game, a foursome on a walking course, several years ago with Gary as my partner. As we were playing for money as well as fun, we were somewhat motivated. On my first hole, I had 18 strokes. On the 11th hole, I nearly birdied … thanks to Gary's helpful feedback, tips and encouragement.

Choose your coaches from those you work with and respect, and get all the feedback and coaching you can get. And, here's a somewhat surprising tip from our learnings about human behavior: sometimes people who are good at something (like sales or management or teamwork) really don't know how they do it! So, you may learn more from **watching** them than from asking them. (More about this in the next chapter.)

Tip: Use your sensory acuity and observe them carefully, listen carefully, notice patterns of movement or expressions or gestures, etc.—not once but several times. Make a "movie" of how they do it.

Think about "coaches" you use now (whether they know it or not) as well as some coaches you might use in the future:

Coach (or potential coach)	Skill or Ability
____________________	____________________
____________________	____________________
____________________	____________________
____________________	____________________
____________________	____________________
____________________	____________________

Coach (or potential coach) *... continued ...*	**Skill or Ability** *... continued ...*

Chapter 6

LEARNING HOW TO LEARN

As we head into the next millennium, global economic competition with the need for everything to be better, faster and cheaper, something else is also happening. A fundamental shift is taking place regarding **learning** and how, when and why we pursue learning.

> *In times of drastic change, it is the learners who inherit the future. Those who have finished learning find themselves equipped to live in a world that no longer exists.*
>
> ... Eric Hoffer

The so-called "half-life" of professional knowledge (the period in which half of what we need to know is either new or obsolete) grows shorter for all professions. As a consequence, **learning** becomes a survival skill for **everyone**.

> *For an organization to survive, it must have a level of learning that minimally equals the level of change in its environment ... and for an organization to learn, individuals must learn. What do these thoughts mean to you? For you?*
>
> ... Sharon Gilchrest O'Neill
> ***Lurning***, Ten Speed Press, Berkeley, CA, 1993

In ***The Seven Habits of Highly Effective People*** (Simon & Schuster, NY, 1989), Stephen Covey puts his "sharpening the saw" habit seventh. I'm glad it made the list, but perhaps it should be first on the list. Covey believes that learning should be a **habit**, that is, an activity we use often if not continuously!

But, we have a big problem here! Very few of us know how we learn at all, let alone how we can learn *best.* Because each individual processes information in a unique way, we also learn in unique ways.

So ... How Do We Learn?

Think about several recent learning situations you faced. Perhaps a new piece of computer software; a new piece of household equipment, like a coffee maker or a VCR; or a new procedure or practice at work. **How did you approach it?**

- Did you try to **chunk** it down into single steps or behaviors?
- What did you do **first**, second, next and so on?
- Was that a **way** that you often approach learning?
- Is there possibly an easier or **better** approach for you?
- Perhaps a new step or a different sequence?

> As an example, I've studied my own learning style while asking client groups to identify theirs, based on our real experiences, not our "theories." My own **actual** learning approach is often as follows:
>
> - **I look** over the situation, the aids or equipment, and so on.
> - Based on previous experience, I **figure** out what to do first.
> - **I try** it out and see what happens. Often it works. Sometimes not.
> - **Only** in failure or frustration do I finally consult any guides or directions!
>
> There's an element of risk involved in my approach ... and it is **not** particularly recommended! I broke the choke on a new chain saw with this approach and, only **recently**, my friend Mert discovered a gas gauge on that chain saw that I **didn't** know was there!

This point is worth repeating: Most of us don't know how we learn anything, and **many of us can improve our learning process ... and should**! But first, take the opportunity to think about and identify **how** you learn:

____

____

____

____

____

____

Learning More Easily

The many behavioral scientists who have contributed to the development of NLP (Neuro Linguistic Programming) have studied learning in more depth than perhaps anyone else. Through an activity that they call "modeling," they pay little attention to what people say they do and more attention to what they **actually do** to complete some difficult job or challenging activity, such as skiing, shooting, swimming, or singing.

Richard Bandler and John Grinder, two of the original developers of NLP (***Frogs Into Princes***, Real People Press, Moab, UT, 1979), make a strong claim. If **any** person can do something, so can **you**. That is, you can learn by "modeling" what another person does, including the parts that even they don't know they do or don't know how they do.

Here are some **helpful tips** that will enable you to learn a skill or an activity faster, easier, more effectively, with more fun and less frustration:

- If you possibly can, watch skilled people **do** what you want to **learn**. Watch them several times and from different vantage points.
- If possible, ask them to "slow it down" so you can see the steps or chunks more easily as they do the activity.
- Also, be mindful of the **sequence** of the steps or chunks, including their mental (thinking) steps!
- And, if you can, interview them as to their **beliefs and mental** processes which you may not be able to see.
- Recognize that most skillful activities have a most effective **strategy**.

> For example: for many people, the best **spelling strategy** is to look upward (typically to your left) to "see" the word in your mind, then to look downward (typically to your right) to "sense" if it feels right.

Tips for Special Situations

- If you become frustrated by a learning effort, **relax** and put yourself into a resourceful state of mind and posture. Then, try again.
- Also, when frustrated, try approaching the situation as a **puzzle:** be curious and stimulated; find the hidden key to a breakthrough!

- If the learning activity is very important to you, find a good **coach** who will help you discover whatever you're missing.

- If the activity is **risky** in some way to you (ex: diving) or to your organization (ex: hiring), be **sure to get an excellent coach!**

I recently took my own advice (imagine!) while resetting our new programmable thermostat. It has 11 function buttons that read out in 14 locations with very tiny print, poorly lit. Holding the directions in my left hand and my flashlight in my mouth, I was thrilled to master it in 15 minutes. (Of course, my *forgetery* will require that I relearn it next spring.)

Making Use of "Mistakes"

It's important to note that learning some things is **risky business** and it is therefore advisable to **load the situation for success** by means of guidance, coaching, safety precautions and equipment, etc.

That said, we need to redefine "failure" in many situations as simply a challenge and a learning opportunity. For example, Thomas Edison tried hundreds of materials for a light bulb filament that didn't work. He regarded that as **feedback**, not failure.

Many philosophers over the ages have suggested that **learning and gaining wisdom** is one of the key parts of life, if not the central part. Our mistakes, failures and faux pas have an important role in that process of learning.

A strong argument for this is that, in my own experience, I seem to make as many "mistakes" at 64 as I did at 16. Usually, but not always, they are different mistakes. (I seem to be a slow learner, but I am teachable!)

Something that helps me maintain a positive perspective and learn from failure is to view that effort or money as simply "tuition"—the cost of learning. Some of my recent examples include:

- An unwise equity investment choice I made.

- Overly frank feedback I gave a client executive.

- Fixing a heat pump I should've replaced, and later did.

Personal Use of the Shewhart Cycle

Many individuals, teams and firms have made excellent use of the **Shewhart Cycle** (***The New Economics***, Second Edition, by W. Edwards Deming, published by MIT Center for Advanced Educational Services, Cambridge, MA, 1994) for continuous improvement in their processes.

The Shewhart Cycle is very appropriate for use by anyone in his/her daily living and learning and, in its simplest form, looks like this:

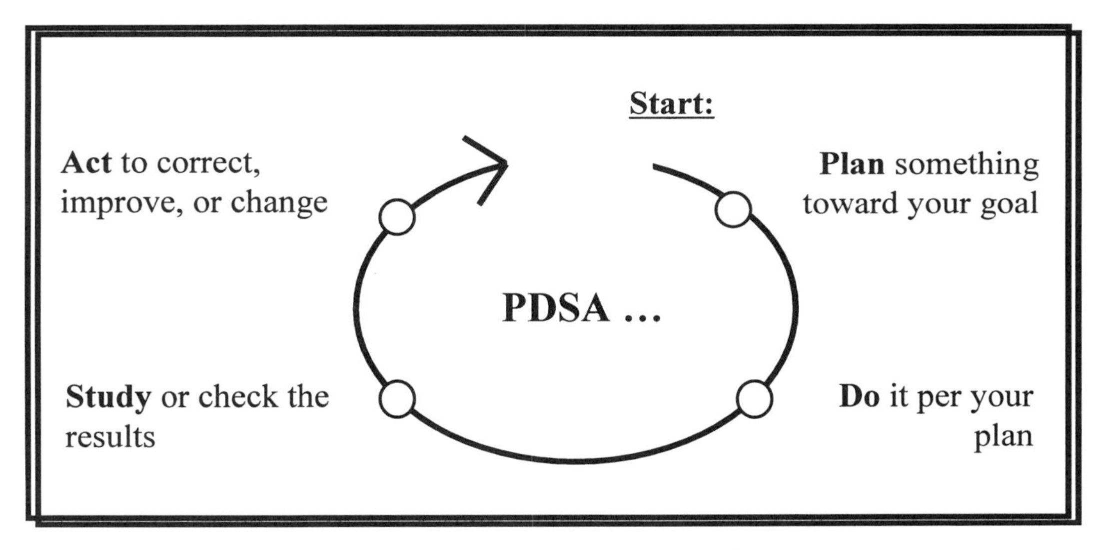

Figure 6.1: The Shewhart Cycle

Particularly the first time we try to do something, in spite of careful planning, we are likely to experience **unexpected consequences**. There are often surprising, mysterious, unpredictable consequences.

The more complex the activity or situation, the more likely there will be an unexpected outcome. And sometimes, it is **nasty**!

> For example: I was recently sawing a dead limb off a large, tall tree. The limb was on the side opposite my ladder. The limb took a bizarre bounce and fell heavily against the ladder! Fortunately, I had tied the ladder top to the tree trunk, a learning from a previous nasty learning experience.

Other Learning Tips

Following is a list of learning considerations that people in all fields of work may find useful or helpful:

- **Ask lots of questions.** Questions without assumptions, judgments, accusations, or statements in disguise. A good question is a super tool we can use to learn almost anything.
- **View your organization as a collection of learning resources**, and use it to:
  - study role models
  - seek out coaches
  - learn from customers
  - talk with suppliers and vendors
  - consult with experts (inside or outside)
  - check out libraries, databases, videos and other visual resources.
- **Seek feedback on how you are doing** from:
  - your own group
  - internal customers
  - external customers
  - suppliers and partners
  - your subordinates
  - your manager or board
  - respected competitors
  - mentors and coaches.

> *I never let my schooling interfere with my education.*
>
> … Mark Twain

Chapter 7

APPRECIATING DIFFERENCES IN PEOPLE

[This chapter is an amended version of Chapter 10 from my first book for ASCE: ***Collective Excellence***, published in 1992. It's been reshaped to support the purpose of **this** book.]

It's been my privilege to assist many management groups that are highly multicultural and diverse in their membership. They've included American minorities, women, and people of wide experience, education, and age span. More recently these groups have been of mixed nationalities, including European and Asian managers as well as American. As challenging as these groups are for their consultant, they are even more challenging for one another!

A critical skill for all kinds of careers in the years ahead, will be that of understanding, appreciating and working well with colleagues, customers and partners who are quite different from ourselves.

These differences will be many, including

Δ country of origin

Δ educational background

Δ culture and traditions

Δ working experiences

Δ personality types.

Diversity and Strength

When group members appreciate their differences, these diverse or multicultural groups have several strengths and advantages that may materialize in their work together. When other factors such as education and experience are equal, these groups are generally more lively, creative, and wide-ranging in their thinking than less diverse groups.

A word about "culture": **Culture is the sum of "how we do things around here."** It's influenced by traditions, myths, history and heritage. Culture shows up in norms of behavior (unwritten rules) as well as in rewards and sanctions.

Coming from a wider background of cultures, subcultures, and traditions, members of diverse groups have more divergent perspectives. They are less likely to take things for granted and will probably have a broader base of knowledge.

This strength, however, may be and often is also a weakness. This broader knowledge base, clash of norms, not taking things for granted, and challenging one another's thinking can make members uncomfortable with one another.

A very basic need here is a tolerance for differences in people and their perspectives. But **more than tolerance** is required for high-performing teams. Members must come to **appreciate** their differences for the value they bring to the team and in spite of the discomfort that may come with it.

Personality Differences

Beyond differences in culture, norms, and perspectives among members, there are some other important differences among people that must be recognized and appreciated to be effective in one's work and life.

Recognition of personality type differences goes back many centuries. Hippocrates spoke of the four temperaments in 400 B.C., which he identified as:

- Δ ***sanguine:*** the extrovert, optimist, talker
- Δ ***melancholy:*** the introvert, pessimist, thinker
- Δ ***choleric:*** the extrovert, optimist, doer
- Δ ***phlegmatic:*** the introvert, pessimist, watcher

Most people have aspects of **several** personality or temperament types, which adds both spice and complexity to the task of working together.

Members of groups owe it to themselves and to their colleagues to develop a good understanding of personality or temperament types. Many excellent resources have become available in recent years. Here are some of my favorites:

Δ ***What Type Am I?: Discover Who You Really Are*** by Renee Baron, published by Penguin Books, New York, NY, 1998.

Δ ***Type Talk:*** *Or How to Determine Your Personality Type and Change Your Life* by Otto Kroeger and Janet M. Thuesen, published by Delacorte Press, New York, NY, 1988.

Δ ***People Types & Tiger Stripes:*** *A Practical Guide to Learning Styles* (second edition) by Gordon Lawrence, published by Center for Applications of Psychological Type, Inc., Gainesville, FL, 1982.

Δ ***Personality Plus:*** *How to Understand Others by Understanding Yourself* by Florence Littauer, published by Power Books, Fleming H. Revell Co., Old Tappan, NJ, 1983.

Δ ***Gifts Differing*** by Isabel Briggs Myers with Peter B. Myers, published by Consulting Psychologists, Inc., Palo Alto, CA, 1980.

All but one of these books are based on the basic personality types identified by Carl Jung and serve as excellent companion books. Littauer's book is more basic, but provides excellent background material for any other reference or method of understanding people differences.

Other Differences

Many other systems, profiles, and typologies exist for noting differences in various important aspects of working together, such as:

Δ styles of leadership

Δ approaches to conflict

Δ methods of problem solving

Δ ways of processing information.

Any of them is quite useful in appreciating and using the natural differences in people. Some such differences are important because they frequently pop up and often cause irritation or annoyance among colleagues, partners or customers.

Some of the most frequently observed, heard, or felt differences among people working together are shown in *Figure 7.1*. They are in no particular order and there is no intended correlation between the left or right side of one aspect and the left or right side of another.

Internal, own goals and standards	----**Locus of Control**------	*External, others' goals and standards*
Preference for being with other people	--------**Sociability**---------	*Preference for being alone or with one person*
Likes details and sensory data	--------**Perception**---------	*Likes imagination and concepts*
Tends to trust logic and standards	---------**Decisions**----------	*Tends to trust values and feelings*
Decisions, plans, and schedules	----**Copes by Way of**------	*More information, options, flexibility*
Style, flair, appearance	-------**What Counts**--------	*Substance, content, value*
Conflict is OK, let's argue	-------**Disagreement**-------	*Be harmonious, don't fight*
Reactive, responsive to others	-----**Source of Action**-----	*Proactive initiatives and actions*
Work, work, work	----**Preferred Activity**----	*Take time to enjoy*
Do it quickly	-------**About Work**--------	*Do it well*
Be calm and cool	----**Showing Emotion**-----	*Let it all hang out*
Pessimistic, cautious	--------**World-view**---------	*Optimistic, enthusiastic*
People	--**Focus of Management**--	*Tasks*
Directing	--------**Manages by**--------	*Leading*

Figure 7.1: Ranges of differences in people noted frequently in organizations and groups (No significance of placement of left/right sides)

Some are normally distributed in the general population. Others are skewed to one side or the other. In some organizations, **types tend to cluster** because of the selection and/or training and enculturation processes in the organization (or subunit of it).

Racial, Cultural, and Gender Differences

While significant progress has been made in understanding and bridging these differences in North America (to varying degrees), much work remains to be done in almost all organizations.

Stereotyping is a natural human tendency, even helpful sometimes in that lumping groups of things or events or situations together can simplify living a great deal. It would be terribly wasteful if we had to learn again each time that rocks are hard, fire is hot, and so on.

> **Problems** that arise from stereotyping are the following: (1) We often apply our skills at categorizing and lumping things together to people who are only superficially similar. (2) We accept what others have said about groups without using facts readily at our disposal. (3) When we do look at what is true about individuals, we consider them to be "exceptions to the rule."

These last three items constitute the practice of prejudice, or pre-judging (based on superficial or inaccurate data, assumptions, or hearsay). Author Rob Terry notes that the power to act or influence others (to exclude people, to hire or not hire, to pay well or poorly, to promote or not) **coupled with prejudice** equals racism, sexism, and other acts of discrimination, like so:

(Power x Prejudice) = (Racism, Sexism, Ageism, etc.)

More on Stereotyping and Prejudice

Most of us are **not** aware of the hundreds of subtle myths, half-truths, beliefs, and nontruths we have inherited (unexamined) from the important people in our lives, including parents, grandparents, siblings, play and school mates, and the media. Even our churches, schools, and agencies have contributed (perhaps unwittingly) to these learned prejudices.

Gordon Allport's studies of prejudice development showed that 70 percent of the people studied attributed learnings about prejudice primarily to parents. He found that such prejudice in children was learned considerably later in life than when children learn most other important life attitudes (from 0 to 6 years of age).

He also observes that, once learned, we tend to keep our prejudices and stereotypes **intact** by:

Δ **Lack of data:** Whether through lack of experience with people or our own internal filters or both, we simply do not "see" that individuals are not like we stereotype them to be. The evidence goes unobserved or undigested.

Δ **Exceptionalizing:** We believe that specific individuals we know are "exceptions," i.e., they are not lazy, incompetent, or overemotional, like the "others out there."

Δ **Re-fencing:** When finally confronted with massive data about either individuals or a group that is contrary to our prejudice, we may fall back on some other aspect of our prejudices.

It is in this area that we are most likely to misuse our intellectual abilities to kid ourselves. We must be willing to "hear" feedback, both verbal and nonverbal, to gain the needed knowledge in spite of our biases.

The research on IQ mentioned in an earlier chapter also suggests that there is a wide range of talents and abilities among members of every group, and no group is inherently superior to others. It's also important to realize that valuable contributions to humanity have been made by persons from both genders and all racial, ethnic, religious, physical, and socioeconomic backgrounds.

Dealing with Differences Effectively

It's easy to be prescriptive about any differences in people, to believe that one aspect is "right" and another is "wrong." But this is not necessarily so. It seems to take all types to make the world work, no matter how uncomfortable these differences may feel to us.

Psychology, behavioral science, and management studies have given us some clues about how best to deal with these and other differences in people:

Δ **Recognize that diversity** in a group not only adds to the conflict potential, but also adds to the group's potential creativity with better solutions and the individual's potential for learning and development.

Δ **Consider the possibility** that when another individual is making you uncomfortable (though not in any way threatening to you), it may be due to their challenging some belief, opinion, bias, or prejudice of yours, giving you a new perspective or new information, or perhaps reminding you of something you've disliked in yourself.

Δ **"Hang in there"** in situations where differences are difficult for you, particularly if you have a sense of progress. Consider asking for outside counseling or facilitation if that might help. Be open to feedback from the other party/ies.

Δ **Put yourself in their shoes**; try to experience their side of the situation. How does it sound, look, feel now from **their** perspective?

Δ **Mentally move away** and try to become a neutral observer or third party (become your own consultant). How does it sound, look, feel now from that perspective?

Δ **Change their behavior** toward you by changing some of **your** behavior toward them. No one can directly change another person. However, individuals will certainly change in response to a change you make, if the change is positive and visible (to them) and it is held long enough to be noticed and "tested" for genuineness.

An example: Sally, vice president of marketing, was often confused and upset whenever CEO Ralph reviewed her plans or work critically. Sally made two changes. First, she now asks Bob (vice president of human resources) to critique her work before her meetings with Ralph. Second, she now notes on a pad each of Ralph's points as he makes them.

As a result, Sally is less apprehensive (before) or upset (during meetings) and Ralph is more respectful in his criticism (in response to Sally's changes).

Helpful References

Δ ***The Nature of Prejudice:*** *A Comprehensive and Penetrating Study of the Organization and Nature of Prejudice* by Gordon W. Allport, published by Doubleday Anchor Books, Garden City, NY, 1958.

Δ ***The Content of Our Character*** by Shelby Steele, published by St. Martin's Press, New York, NY, 1990.

Δ ***For Whites Only*** by Robert W. Terry; William B. Eerdman's Publishing Company, Grand Rapids, MI, 1974.

Things to Think About

Δ **What do you know about your personality type(s) from various profiles or feedback?**

Δ **Strengths of your personality:**

Δ **Potential weaknesses to be mindful of:**

Δ **What are some stereotypes you notice in yourself and others from time to time?**

Chapter 8

UNDERSTANDING EMOTIONS/FEELINGS

About Emotions

It is sometimes difficult for us to accept that feelings and emotions **are** such an important part of life, and something we **must** deal with in others and in ourselves.

This is especially true when feelings (emotions) are untimely, unseemly, confusing, inconvenient, upsetting, distracting and unwelcome.

Webster defines **emotion** as a state of strong **feeling** (such as anger or fear) experienced both psychically and physically. Here is a list of emotions or feelings we may expect to experience in life:

😐 afraid
😐 aggravated
😐 agitated
😐 alarmed
😐 amused
😐 angry
😐 annoyed
😐 anxious
😐 apprehensive
😐 aroused
😐 astonished
😐 bitter
😐 calm
😐 comfortable
😐 concerned
😐 confused
😐 contented
😐 cross
😐 dejected
😐 delighted
😐 depressed
😐 disappointed
😐 discouraged
😐 disgruntled
😐 disgusted
😐 dismayed
😐 displeased
😐 distressed
😐 distraught
😐 disturbed
😐 downcast
😐 downhearted
😐 ecstatic

- 😐 elated
- 😐 electrified
- 😐 embarrassed
- 😐 enthralled
- 😐 exhilarated
- 😐 frightened
- 😐 frustrated
- 😐 furious
- 😐 glad
- 😐 grateful
- 😐 grieved
- 😐 happy
- 😐 horrified
- 😐 hurt
- 😐 infuriated
- 😐 irked
- 😐 irritated
- 😐 jealous
- 😐 jittery
- 😐 joyful
- 😐 joyous
- 😐 jubilant
- 😐 lonely
- 😐 mad
- 😐 melancholic
- 😐 merry
- 😐 miserable
- 😐 mortified
- 😐 nettled
- 😐 overjoyed
- 😐 pleased
- 😐 rancorous
- 😐 relieved
- 😐 resentful
- 😐 sad
- 😐 scared
- 😐 shocked
- 😐 sorrowful
- 😐 spell-bound
- 😐 splendid
- 😐 surprised
- 😐 taken aback
- 😐 tense
- 😐 terrified
- 😐 touched
- 😐 tranquil
- 😐 troubled
- 😐 undone
- 😐 uneasy
- 😐 unhappy
- 😐 upset
- 😐 vexed

Emotions have been an almost overwhelming part of my life at times, particularly for my first 35 years. My emotions seemed to drive almost everything I did.

Many of those emotions were what can be described as unpleasant or negative—feelings I'd rather not have had. Anger, frustration, hurt, fear, and impatience were my **frequent companions** as I tried to cope with …

- 😐 alcoholic parents
- 😐 an assertive spouse
- 😐 a demanding job
- 😐 four active children.

My Journey of Discovery

Some of my relatives, friends and work colleagues seemed to have emotional strength, self-confidence, poise, contentment and most of all … a lot of happiness. It gradually dawned on me that they somehow generated these feelings, **internally**, for themselves, no matter what "happened" to them.

And so, I began a lifelong effort to …

- ☺ better understand my own emotions
- ☺ become less of a hostage to them
- ☺ better understand the emotions of others
- ☺ gain more contentment in my life
- ☺ be a better spouse and father
- ☺ be more effective at my work.

I pursued quite a few personal development experiences, workshops, and conferences and gained insight from the counselors that I met there.

The most useful learnings were experiential—learning through experiences, sometimes painfully and sometimes playfully.

Some of these experiences were incredibly helpful and had powerful, positive impacts. Others were so-so and a few were awful!

Feeling "Types"

Following are a few of the **emotional** "facts of life" that I discovered in my 30-year journey and quest for learning about feelings and emotions:

- ☺ Some people are very comfortable with their "feeling" side and use their emotions as valuable **information** to consider, along with other important data.
- ☺ Others are pretty well **controlled** by their emotions, which seem to direct their behaviors, decisions and lives. (I was here.)
- ☺ Still others appear to be **out of touch** with their emotions, (seemingly) using logic alone to guide their lives. (Later I learned this last group is also influenced by emotion, but perhaps below their level of conscious awareness.)
- ☺ And some people use their emotions to attempt to **control others**, whether consciously or unconsciously.

From all of this, it seems most useful to be **aware** of our emotions as clues to our perceptions, values and expectations. And, it is also useful not to be controlled or "run" by our emotions or those of others!

Unscrambling Emotions

A "model" I've built for myself helps me unscramble what's happening whenever I find myself in an emotional swamp or morass (see *Figure 8.1*).

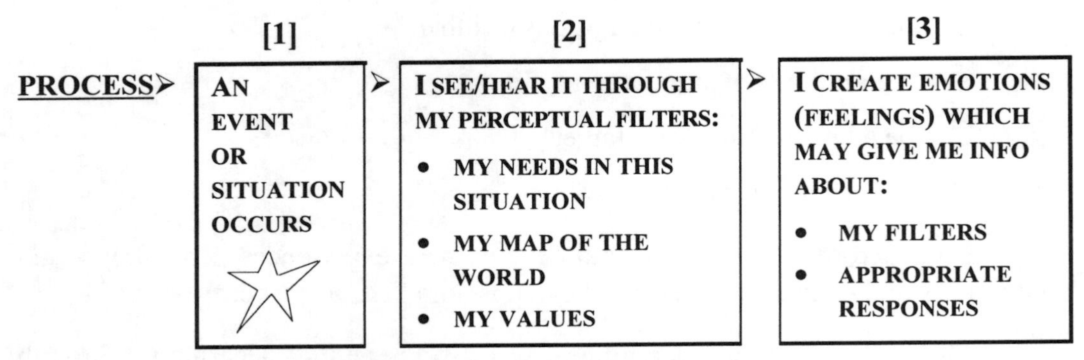

Figure 8.1: A "Feeling" Process

This process (or something like it) happens so quickly (probably in nanoseconds) that we may be unaware of part [2] in this sequence.

Some of our reactions are almost always **in**voluntary. For example, if I feel uncomfortable in a situation, I may flinch, flush, frown, tense my body, etc., perhaps without even realizing it.

And sometimes, the **trigger** for a feeling is below my level of conscious awareness, such as a fragrance, a background noise, a tone of voice, a familiar scene, or a slight touch.

Needs and Expectations

Many of us go through life believing that our own feelings are caused by **others and external events**—something outside ourselves. This usually isn't true and can be a disempowering belief in any case.

The second part is the most critical: It's disempowering to believe our emotions are caused by others because it provides the opportunity for **others and external events** to manipulate us!

We often have one or more powerful subconscious **needs** as a result of our upbringing and very early life experiences. Here are some common subconscious needs that may disempower us if we let them:

- 😐 Need for **approval** of others
- 😐 Need to always be **"right"**

- ☹ Need to be **self-sufficient**
- ☹ Need to always be **"strong"**
- ☹ Need to be **loved/lovable**
- ☹ Need to **get it all done**, etc.

These irrational needs create expectations for ourselves that are unreasonable and inappropriate! They also create many of the feelings we experience.

These irrational needs may also lead to expectations of others that are neither realistic nor appropriate. To paraphrase Fritz Perls, we are not in the world to meet one another's expectations. But if it happens, consider it wonderful!

Developing Emotional Balance

In my growing up years, most of us were taught that feelings generally were not to be honestly expressed in public unless they were socially acceptable. (And **few** feelings were!)

The trouble is, those feelings "leak" out in funny ways anyhow! Those leaked feelings often confuse people and create nasty problems in relationships, in communication, or in mental health or physical health.

Then, not too long ago, pop psychology books and T.V. shows suggested it was important to "let it all hang out" emotionally. Some of this was good; for example, some men began to allow themselves to cry and show their caring side. I tried it and learned from the experiences. But, I don't recommend "letting it all hang out" as a general rule.

Somewhere in the **middle** seems to be a healthier, saner, more productive place to be.

LET IT ALL HANG OUT!	**BE AWARE OF YOUR FEELINGS AS USEFUL INFORMATION**	**BOTTLE UP, HIDE OR BE UNAWARE OF YOUR FEELINGS**

Another interesting aspect of feelings is that often people really **trust** someone who is aware of his/her own emotions and who shares them honestly as **information** without necessarily "acting them out." This is particularly true for hard-to-take feelings like anger, upset, frustration, impatience, confusion, embarrassment, strong disagreement and such.

About Tears (Crying)

One of my most important learnings about feelings came from mentor and colleague Kaleel Jamison … important because tears are so often misunderstood.

Kaleel taught clients that **tears** simply mean that the person is experiencing a very **strong emotion**. It could be joy, sadness, pain, longing, love, anger … anything.

She believed we should be mindful that tears represent a sort of "sacred ground" in that we are being allowed to be present with that person in that moment.

She suggested that we …

- 😐 Not treat them as helpless or incapacitated
- 😐 Not discount their feeling **or** message
- 😐 Listen carefully and respectfully.

Summary of Emotions

Following are some "notions on emotions" that have been helpful to me in recent years…when I remember to use them. Perhaps they'll be helpful to you:

- 😐 Be aware of **your own** emotions as clues to your perceptions, values and expectations.
- 😐 Don't be "run" (controlled) by your emotions; use your other faculties as well; act appropriately.
- 😐 Be aware of **others'** emotions, as clues to their perceptions, values and expectations.
- 😐 Try not to be "run" (controlled) by others' emotions, but use the **data** they provide; act appropriately.
- 😐 If you don't like the emotions you're experiencing around other people, you may want to consider and change one or more of these:
  - your behavior(s)
  - your expectations
  - your associations.

Really Helpful Work by Others

1. Richard Bandler's humor and metaphoric expression is both entertaining and helpful as he shares his thoughts on feelings and other parts of our "human operating system" in ***Using Your Brain For a Change*** (Real People Press, Moab, UT, 1985).

   Richard notes some of us have better vacations in our imagination than we do in reality due to our ability to be disappointed! Not satisfied with that, we even do reruns so we can experience our lousy feelings all over again!

   Any good book shop will order his book for you, as well as the next two books.

2. Mark D. Youngblood gives a powerful presentation in his book ***Life at the Edge of Chaos***. His section on leadership and personal mastery is a gold mine for feelings and more (Perceval Publishing, Dallas, TX, 1997.)

   Part three, Powerful New Ways of Being, provides excellent help on communication, relationships, dialogue and self-mastery.

3. Leslie Cameron Bandler tells of her own struggle with emotions and living with them, offering practical insights and excellent tips for others in her incredibly helpful book ***The Emotional Hostage*** (FuturePace, Inc., San Rafael, CA, 1986).

4. Robert Cooper and Ayman Sawaf believe that emotional intelligence is a learnable capability, one that can be developed at any age. Their book ***Executive EQ*** is published by Grossett/Putnam (1997) and is also summarized on tape by Audio-Tech Business Book Summaries (800/776-1910).

5. Dr. Hendrie Weisinger offers a strong argument that emotional intelligence is probably the major factor governing success at work, for individuals and their organizations, in his book ***Emotional Intelligence at Work*** (Jossey-Bass Publishers, San Francisco, CA, 1997).

Insights for Myself

😐 **What emotions (feelings) do you have often or frequently? And, in what typical situations?**

Frequent Emotions:	Typical Situations:
______________	______________
______________	______________
______________	______________
______________	______________
______________	______________
______________	______________
______________	______________
______________	______________
______________	______________
______________	______________
______________	______________
______________	______________
______________	______________

😐 **If you're not happy about some of these, are there some behaviors, expectations, associations you might consider changing?**

____

____

____

____

____

____

____

____

____

____

Chapter 9

COMMUNICATION ESSENTIALS

Everywhere, we see signs that communication is in distress. The "Money" section of ***USA Today*** (September 30, 1998) reports a significant **decline in the effectiveness** of business communication due to workplace stress, time pressure and **compression of communication**. In trying to do it faster, we're really doing it less well!

Mental Maps

Covered in Chapter One, an understanding of **mental maps** forms the essential foundation for **any** kind of human communication.

Being **un**aware of the differences in our individual mental maps leads to needless confusion, misunderstandings and conflict.

On any important topic, it's well worth the time to explore and explain how **you** feel and what you ... think
believe
want
propose

... while also finding out what the **other person** thinks, believes and wants.

Listening and Expressing

Those who are often considered to be "good communicators" ... self confident, articulate, extroverted and expressive ... need to devote more time, attention and energy to **listening**. **Listening**, according to Semanticist Hyakawa, is potentially the most powerful part of communicating!

Those who are often considered to be "poor communicators"—quiet, soft-spoken, introverted and analytical—need to devote more attention and energy to **expressing themselves**. What you don't express in clear and timely ways **can** hurt you, your colleagues and your organization.

"Non-verbal" Communication

Much has been made of the need to use gestures and understand "body language" in communication. But the **more subtle** forms of non-verbal communication such as posture, voice tone and inflection, eye and eyebrow movement, and rhythm are quite important.

Recently, I was assisting two executives in their approach to a common problem they needed to resolve **together**. **Something each did made the other uncomfortable** as they tried to communicate on difficult issues. Each felt the other was "confrontational" or "uncooperative," but they weren't clear about **why** they felt that way. Observing them closely, I noticed ...

- Jack tended to have a "crooked smile or smirk" and a chuckle at certain points he made. (Interpretation: "Of course this is so; anyone can see that!")
- Michael tended to use a "hard, direct stare" with an edge to his voice at certain points he made. (Interpretation: "This is the way I've decided, like it or not.")

Once they were made aware of these communication mannerisms, they could easily change them. But first, they needed feedback!

Whole Body Listening

In her very helpful book ***NLP at Work*** (Nicholas Brealey Publishing, London, 1995), Sue Knight looks at **non-verbal listening**. She calls it "whole body listening," which has three important functions:

- It helps the **speaker** to know s/he is being really well listened to!
- It helps the **listener** to actually do a far better job of getting the speaker's full message as s/he listens.
- It helps the **listener** to better understand how the speaker feels (his/her "state") and how s/he sees the world (his/her "mental map").

Whole body listening can be briefly summarized as:

- being in a state of **curiosity**
- totally **focused** on the other person
- **gazing** at the other person
- **matching** the other person's posture
- when responding, speaking to the **other's issue**, not to something else
- using key words and patterns, voice level and tone, that are **similar to** those of the other person.

Communication Simplified

John Parmater, a consultant and colleague from Columbus, Ohio, boiled down weeks of NLP training content to the essential and simplest steps for successful communication. Here's what John extracted for "communication simplified":

Know what you **want** ...

- Every time you even think about communicating, it's because you **want** something.
- Be fully aware of what that want is because there may be **many** ways to achieve it.
- Consider **how** you want to communicate what you want or need **before** you proceed.

Know if you're **getting** what you want ...

- Sometimes it's obvious that you are or aren't, by way of the other person's **responses**.
- Oftentimes, it's **not so clear** because you may be getting little information or perhaps mixed messages.
- **Sensory acuity** then becomes essential, so **you** can perceive more communication (data) than most folks do.

- Sensory acuity means paying close **attention** to body posture and gestures, voice tone and pitch, words spoken and deleted, eyes and eye movement, skin tone and more. **People cannot not communicate**.

> If you're **not** getting what you want, do **anything** else …

- **Don't** do more of the same "harder." This simply reinforces the resistance of others.

- Be aware of their **personality preferences** (analytical, expressive, amiable, driver) or Myers Briggs types. Then **complement** those preferences in order to build rapport.

- Be aware of their "state of mind." **Put yourself** into their situation, mentally and emotionally. Mirror their posture and rhythm. See what that **reveals to you** about them, their needs and so on.

- If you don't already know, find out what **they** want or need. Sometimes it may surprise you. They are more interested in what **they** need than what **you** need.

- Start over again. This new **perspective** will suggest many different approaches to getting what you want.

  As you maintain rapport (eye contact, pacing, posture, energy level, etc.), you build trust and respect. This does not mean you necessarily agree with them!

- **Don't** make assumptions as we often do. For example:
  - Silence does **not** mean agreement.
  - Explosive anger may **not** be about you.
  - A question is often really a **statement in disguise**.
  - People are **often indirect** about their wants and needs. So, find out exactly what they feel, need or want.

Take 100% responsibility for a successful communication!

Chapter 10

DEALING WELL WITH ADVERSITY

Scott Peck's best seller ***The Road Less Traveled*** (2nd Edition, Touchstone Books, 1998) begins ...

Life is Difficult ...

... and so it seems to most people at some or many parts of their lives.

I'd decided that an appropriate metaphor for this chapter could be mountains and mountain climbing. I grew up in the mountains around Beckley, West Virginia, a town so steep that it had no airport until recently and so high that it still has no train station.

Mountains are truly beautiful. And they can certainly wear you down! But mountains have also helped me to develop good perseverance in the face of most kinds of adversity. And, I'm grateful for that.

Recently I discovered Paul Stoltz and his practical book ***Adversity Quotient*** (John Wiley & Sons, NY, 1997), including his Adversity Response Profile (ARP). He, too, uses a "mountain" metaphor. He notes that professional success can be compared to a steep mountain to be climbed. Then he identifies three ways people deal with their mountains ...

Climbers **Campers** **Quitters**

- **Climbers**, he says, feel a deep sense of purpose, are motivated to make things happen, and want to keep improving themselves.
- **Campers**, on the other hand, find a "comfortable plateau on which they can hide from adversity," just staying safely employed.
- **Quitters** "spend their careers avoiding risks" (and challenges and difficulties) while contributing little of value.

Attitudes and Adversity

Many studies in the fields of management, psychology, education and social work clearly indicate the importance of our **attitudes** in coping with adversity. Three of my favorite commentaries capture this very succinctly:

- Whether a person thinks he can, or thinks he cannot … he's right.

  … attributed to Henry Ford

- Life is either a daring adventure or … life is nothing at all.

  … attributed to Helen Keller

- Whatever you can dream and truly want, you can achieve.

  … attributed to William James

For hundreds of years, wise and motivating speakers and teachers have been sending us this message, yet few people actually change in response to these powerful (and true) words. **Why?** It would be too easy to simply say that "quitting" or "camping" is a lot less work than "climbing."

> I believe that how we deal with adversity is a function of attitudes that are deeply embedded in our **subconscious**—the operating system of our bio-computer. As such, they are generally outside of our conscious awareness. And so, others are sometimes more aware of our attitudes than we are.

Sources of Adversity Attitudes

From years of work and consulting practice for many organizations and individuals, as well as reading in management and many related fields, and by being a very curious student of people, I've come to the following beliefs:

- Attitudes and skills at dealing well with adversity are **critical** to professional success in most fields, and certainly in business.
- They are somehow developed or learned **very** early in life and become part of our mental and emotional "software."
- These attitudes and skills can be, and often are, either **strengthened** or **weakened** by traumatic events and by those close to us.
- They can also be deliberately strengthened by any individual for him or herself, if that's something s/he **wants**.

Developing and Strengthening

A number of researchers, authors, teachers and counselors have contributed to the following ideas. These ideas for strengthening what Paul Stoltz (see page 51) calls your **Adversity Quotient** are few and simple.

They need to be simple! To be effective, they must help an individual with his or her "operating system"—**the subconscious mind**.

Table 10.1: Strengthening Our Own Ability to Deal with Adversity

PHASE	ACTIVITY(IES)
1. Create a solid base of **personal power**	Take full **responsibility for everything** that has happened, or is happening, in your life. Not "blame," just take personal *response-ability*, perhaps shared with others in your life. This gives you deeper insight and the **power to re-do** some things in your life you don't like at this time.
2. Check your personal **"map and compass"** for guidance	Refer to your **Mission** work done in Chapter Three. If you didn't develop a Mission then, perhaps **now** is a good time ... Identify **several important goals** for your life at this time. Consider these potential aspects of life for goals: • work and career • family and extended family • personal development • spiritual development • avocations and interests

Table 10.1: Strengthening Our Own Ability to Deal with Adversity
... continued ...

PHASE	ACTIVITY(IES)
3. See yourself succeeding; create a **success scenario**	⛰ **Visualize your goals as if completed**, with the outcomes you hope for, as clearly as you can. ⛰ **Write your goals** as if they are completed, as if they are already true. **Example: I love being organized!** ⛰ **Review these goals daily!** Put them on Post-Its, on your desk, wherever you'll see them frequently!
4. Take action and **get going**	⛰ It's very important to take a **first step**, to begin, to move from mental work to actual activity. ⛰ Do something **concrete** toward each of the goals you developed in Phase 3. ⛰ **Here and now** is often the best place to begin. If you need to correct your aim, remember PDSA (Chapter 6).
5. **Celebrate** successful steps and **redirect** as necessary	⛰ **Enjoy and celebrate** actions you've begun, even those that weren't "perfect" (few ever are). ⛰ Where necessary, take **corrective actions** or "whatever it takes" to keep moving in the desired direction. ⛰ Accept any setbacks with **humor**! For example, I often ruefully observe it takes me two to four times as long to do something as my original "plan" for it.

Other Thoughts

Occasionally, I've found myself to be **over**-committed to some goal way beyond its real importance to me or anyone else! This was particularly true in my earlier days of career and family. It reminds me of a clever comment I've heard several times:

> **If at first you don't succeed, quit! No sense being a darned fool about it.**

Assuming one is pursuing an **important** task or goal, a more useful way to phrase this might be as follows:

If at first you don't succeed, stop and think about it. Consider what you've learned so far. Then try again with a wiser approach!

Sometimes finding a "wiser approach" should depend on **others**. That is, seeking others' advice, testing others' experiences, getting professional assistance, asking for help from team members, colleagues, family members and friends!

Pretending to ourselves and others that we "have all the answers" may boost our ego for the moment ... but, in the longer-run, it's simply **self-defeating** in many ways.

As Calvin Coolidge said, **"Press on. Nothing in the world can take the place of perseverance. Talent will not; nothing is more common than unsuccessful men with talent. Genius will not; unrewarded genius is almost a proverb. Education will not; the world is full of educated derelicts. Persistence and determination alone are omnipotent. Press on!"**

Gerry Kushel, an AMA colleague, shares these important thoughts in his wise book: ***Reaching the Peak Performance Zone*** (AMACOM, New York, 1994):

> **You are, in a very real sense, in charge of your own destiny, whether you want to be or not.**

How I Deal with Adversity

In what areas of life (home, hobbies, earner, spouse, parent, church, relationships, and so on) do I tend to be a Climber, Camper or Quitter?

Climber	Camper	Quitter

Am I satisfied with my perseverance at dealing with adversity? If not, what steps might I want to consider?

Chapter 11

LEADERSHIP WITH A SMALL "L"

Why, you might well ask, should "leadership" be part of a book on **personal effectiveness?** Good question. I've come to think of "leadership" as a **skill set**, one that's important to all of us in some situations. Leadership is an **activity** that answers a situational need that **anyone** may rise to meet.

Also, in spite of all the hype and tripe that's been written over the centuries on leadership—or maybe because of that—we still know **far too little** about leadership. So, I'll try to steer clear of clichés and philosophy and offer a few practical approaches.

In this chapter, we'll look at some practical notions and tips from people I believe **really understand** leadership with a small "L" ... including:

- Karl Albrecht ... who wrote ***Personal Power*** (Shamrock Press, San Diego, CA, 1986).
- Judith Bardwick ... author of ***In Praise of Good Business*** (John Wiley & Sons, New York, NY, 1998).
- William Cohen ... writer of ***Art of the Leader*** (Prentice Hall Trade, Upper Saddle River, NJ, 1991).
- Aubrey Daniels ... who wrote ***Bringing Out the Best in People*** (McGraw-Hill, New York, NY, 1994).
- Robert Greenleaf ... author of ***Servant Leadership*** (Paulist Press, Mahwah, NJ, 1977).
- Peggy Morrison ... who gave us **Making Managers of Engineers** (***Journal of Management in Engineering***, published by ASCE, October, 1986).

Mistaken Beliefs

Judging by what often passes for "leadership," there must be some mistaken beliefs about what leadership really is. Here are several mistaken beliefs that have been discovered through our practice as well as by the folks listed earlier:

- Some believe that one must be in a position of authority to demonstrate leadership. Not so. **In fact, the clearest way to identify a leader is to notice those people who lead without authority and get things done.**
- We often think that each organization needs a leader. But actually, **leaders are needed in all parts and all levels of organizations of all kinds.**
- Leadership is not ordering or directing others to do this or that. Rather, it is **inviting others to do something that is appropriate to the situation.**

> *Leadership: The art of getting someone else to do something you want done because he wants to do it.*
>
> ... Dwight D. Eisenhower

- Leaders have clear goals and high expectations. But, **they are also very interested in the goals of other committed individuals, and in synergy, the power of group wisdom.**
- Difficult, dangerous, emergency, painful and dirty work usually isn't sought out by most folks. Yet, **leaders are often able to turn these aspects of such situations into motivating challenges.**
- Leaders do have the ability to solve problems, but not all problems. Rather, **they are able to break problems apart and help focus others on the parts.**
- Effective leaders aren't afraid to make quick decisions. However, **they often prefer to mull them over and consult others when time allows.**
- Americans in particular, expect reasons for requests and directives. **Leaders find simple truth, well said, usually provides enough "reasons."**

- An "open door" policy is a nice start. However, **an "open ear" policy is better.** Most serious problems are well understood by someone in the organization who would be happy to communicate them if someone was ready to listen.

- Leaders sometimes have "charisma" ... or seem to. But what looks like **charisma is often simply taking the mission and people seriously.**

- To lead (with a small "L") you don't need the power to reward beyond simple, sincere **thanks and acknowledgment**.

- Executives usually believe the biggest motivators are job security, high pay and good benefits. Actually, **the top motivators really are:**

  - **doing interesting work**
  - **receiving sincere appreciation**
  - **having real responsibility for their work.**

> *Contrary to the opinion of many people, leaders are not born. Leaders are made, and they are made by effort and hard work.*
>
> ... Vince Lombardi

Appreciation and Acknowledgment

Appreciation and acknowledgment are powerful influences that are parts of "positive reinforcement." The interesting fact is, positive and **negative** reinforcement operate **all the time**, whether we are conscious of that or not! Here are some principles we need to be **conscious** of:

- Work activities that are recognized or even acknowledged, will often be **repeated**.

- Recognition or acknowledgment **need not** be often and **should not** become predictable (as in "always").

- **Specific and descriptive** feedback is more valuable and performance-motivating than general feedback.

- Recognition or acknowledgment **often** needs to be private (one to one) to avoid jealousy or embarrassment.

- Criticism or "negative" feedback usually needs to be private (one to one) for obvious reasons. There are exceptions.

- Extinction (doing/saying nothing) is often effective at **removing** behaviors. This applies equally to behaviors you want and don't want, so don't ignore positive results!

- Criticism, or negative feedback, **is** sometimes necessary. However, it attaches a negative stigma to the giver in the eyes of the receiver, particularly if it's over-used.

- Criticism or negative feedback is most helpful when it gives **all** this information:

  - What happened, as you saw it?
  - What were the impacts on you and others?
  - What would be better next time?

Aspiring to Leadership

There are several predictable things about "leadership" that we notice in our consulting work for organizations of **all** kinds:

- People are often critical of those in formal leadership positions.

- People in formal leadership positions seek little and get little honest feedback from their staffs on "how they're doing."

- Many organizations are constrained in their growth by a lack of enough excellent leaders or potential leaders.

Among others, Peggy Morrison (see page 57) has some sound suggestions for those professionals who aspire to formal leadership roles:

- Take a broader view of your organization, including the needs and priorities of other parts of it.

- Value openness and honesty and learn to use them with tact and gracefulness.

- Polish your skills of listening, negotiating, managing conflict and communicating clearly.

- Learn the group skills of planning and team development, and the individual skills of delegation and follow-up.

Leadership and Authority

There are many kinds of "authority." Here are two of the most important and most misunderstood:

- **Formal authority** is the recognized "right" to lead a group in various activities. Whether acquired from the organization's management or by consensus of the group, it allows you to make certain decisions, instruct members to do various things and to organize and coordinate their activities.

- **Earned authority** is the personal authority accorded to you by the individuals of a group. It is based on your skills, your attitudes, your demonstrated commitment, your work and their trust in you.

We've all participated in situations in which the person with high formal authority also enjoyed high earned authority with group members, and consequently had **a lot of leadership influence**.

And we've probably seen people with high formal authority who so undermined themselves with their interpersonal styles that they developed very little leadership influence.

New Manager

New managers often have understandable fears: fear of failing, fear of looking foolish, and fear of not having the skills to cope with problems.

Too many new managers react to these natural feelings of anxiety with counterproductive ego-building tactics in dealing with people.

Karl Albrecht (see page 57) offers these tips to "new managers," but they are superb advice for **all of us!**

- **Don't fake it**; don't pretend to know what you don't know, and don't pretend to understand what you don't understand; ask questions and learn.

- **Don't confuse your ego with your authority**; use your authority appropriately and without apologizing for it, but use it sparingly and gracefully.

- **Form a mentor relationship** with an experienced manager, a consultant, a management trainer, or a personnel officer. Call on expert advice and think it over carefully.

- **Establish rapport** with the members of your team and keep it, protect it and nurture it. Look out for others and most of them will look out for you ... with tips, cautions, and extra help.

Project Management

In many organizations, people are asked to serve in the role of project management, either part or full time. In many of those situations, this may be an individual's first real leadership opportunity. Also, and unfortunately, the project manager may have **no** "formal authority" and must rely on **leadership with a small "L."**

Project managers (PM) who may lack formal authority need to have personal communication skills and strategies for effectively **influencing their team members** to:

- meet their project objectives
- accept PM suggestions and guidance
- initiate changes in their work
- get and give assistance to team members
- sometimes, improve job performance.

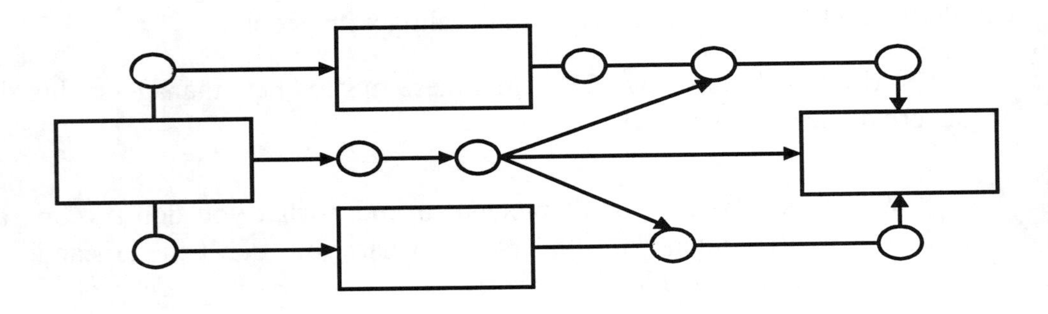

Bohlen, Lee and Sweeney of the University of Dayton studied PMs from the U.S. and Europe and their strategies. They found that those PMs used the following strategies to get results:

1. **Consulting**—Gaining the support of team members by seeking their participation in the planning and decision-making process for a proposal or a project.

2. **Reasoning**—Using factual information about a project to gain understanding, support and assistance for a project.

3. **Relationship**—Building rapport and good relationships with team members before making a request.

4. **Common Vision**—Presenting project objectives in a way that appeals to team members' basic values and interests.

5. **Enlisting Support**—Developing or enlisting the support of other project contributors toward the project objectives in order to influence a particular member.

6. **Higher Management**—Use of higher management authorization and support to directly influence team members.

7. **Asserting**—Using assertive approaches such as clearly stated expectations for assistance to convince team members to do what's needed.

8. **Reciprocity**—Offering an exchange, such as reciprocal benefits for doing what is needed on a project.

9. **Forcing**—Threatening a team member with a negative outcome or penalty if the member does not assist with project needs and objectives.

Different project contributors may and do respond to different strategies, and the effective project manager must be able to use **any of the nine strategies**.

Leadership and You

◎ Does leadership activity appeal to you? When and where?

◎ How could you use appreciation and acknowledgment of others more in your present life?

◎ Whom do you admire for their leadership skills and approach? And, what, in particular, do you admire in them?

Chapter 12
DRIVERS OF BEHAVIOR AND CHANGE

Over the years, I've become more and more suspicious of my own reactions (habits), feelings, thinking and intentions as the basis for important decisions and actions. Each contributes in some way, but each may also contain some pitfalls.

Table 12.1 shows the four things that seem to me to "drive" human behaviors. These four sources of human behavior need to be understood and treated with some caution. All four can be made more useful to us as individuals, as shown.

Table 12.1: Sources/Drivers of Human Behavior

TYPICAL SOURCES OF ...	
(1) Reactions	**(2) Feelings (Emotions)**
• Programmed largely by our subconscious mind; habitual and strongly patterned. (Up to 95% of our daily actions may be programmed this way.) • Often based on "scripts" or "strategies" developed very early in life by our subconscious mind; possibly obsolete.	• Life experiences resulting in pain and pleasure. • "Voices" from our past; both judgment and praise. • Physical traumas associated with pain and unconsciousness. • Emotional traumas and associated pain and stress.

Table 12.1 continued on next page.

Table 12.1: Sources/Drivers of Human Behavior ... *continued* ...

MADE MORE USEFUL BY ...	
• **Noticing** one's own habits and reactions. • **Developing new habits** and "reactive strategies" where appropriate.	• **Noticing** one's own feelings and those of others, but as **data**, not determinants. • Not being "run" or controlled by one's own feelings.
TYPICAL SOURCES OF ...	
(3) Thinking	**(4) Intention**
• The mind's tendency to free-associate and connect things logically and/or creatively. • "One-lesson learnings" (conclusions based on one data point). • The mind's need to decide quickly, without all the facts. • Personal biases, prejudices, myths and stereotypes.	• Largely driven by one's personal values and life goals as they emerge and as they change. • Influenced by reflecting on significant events. • Evaluated in discussions with significant others. • Commitment may be developed sufficient to override faulty reactions, feelings, and thinking.
MADE MORE USEFUL BY ...	
• **Treating** one's own and others' thoughts, ideas and opinions simply as **data**. • Remembering the computer rule: garbage in = garbage out.	• **Keeping** one's relevant goal(s) in the forefront of the mind by reviewing them daily. • Staying focused on what's important in particular situations.

Paradox and Mystery

Once I began to notice how illogical, paradoxical and mysterious my own behavior was (is), I also noticed it in others: family, colleagues, friends and clients. And so I began to "collect" what I might call paradoxes and mysteries of human behavior. Here are a useful few:

Mystery: Our minds hate mysteries or unanswered questions, which create discomfort. In a hurry to get comfortable, our minds "make up" (invent) incorrect answers. This rationalizing process may prevent our questions ever **really** being answered.

Knowledge: Those who know the least are often the most "generous" with their knowledge, offering others unsolicited advice and opinions. Really wise people often have few "answers."

Information: Most of the time, for most of us, giving **information** per se isn't a useful way to cause change. However, if such information is helpful and specific feedback given, with tact and honesty, it **may** result in change.

Control: Control is usually an **illusion**, not a reality. Sometimes we may feel as if we have control, but ultimately, that's a dangerous illusion. (Parents don't even control their children; often it's the reverse.) The best we can usually hope for is influence.

Reinforcing: When a subordinate continues to do something that is not useful or productive, or is wrong, that behavior is probably being **reinforced** by his/her manager's actions or inactions, although probably not intentionally.

Approval: The need or desire to be liked or approved of usually leads to being disliked and disapproved of. People easily sense approval-seeking behaviors and generally respond negatively.

Respect: On the other hand, respect is a result of integrity, honesty, courage and values-in-action. Consequently, respect is more likely to develop out of taking (appropriate) blame for errors than it is from "lookin' good."

Selling: Whether to a client or a colleague, "selling" isn't often done by persuading. It's more easily done by listening carefully to identify needs or problems and then participating in their solution.

Worry: Worry, I believe, is simply a negative goal. Many worries, thankfully, never come true Replacing our worry with a positive goal and focusing on that is a more worthwhile use of energy.

I believe that each of these paradoxical human behaviors is driven by a "faulty" or erroneous part of our mental maps. Yet, **how are we to know** what and where those "faulty" parts are?

Helpful Notions

The following three notions, paradigms, or ways of viewing reality have proven to be very helpful. They can serve us in almost all situations involving other people, one at a time or in groups.

- The first notion addresses **"knowing what you know and don't know"** with several watch-outs.
- The next notion proposes that you focus on those things you can **actually change or at least influence**.
- The third notion suggests that you **really can change other people**. But it requires something special.

A. Know What You Know and Don't Know

With a little help from Mark Twain, Werner Erhard and several colleagues, I've cobbled together what might well be called the "What I Know Chart" (*Table 12.2*):

Table 12.2: What I Know Chart

KNOW THAT I *KNOW*	KNOW THAT I *DON'T* KNOW	THINK I KNOW, BUT *DON'T*	DON'T KNOW THAT I DON'T KNOW
• I can act on this knowledge … • But I need to be very vigilant for indications that I'm **wrong**.	• I can get information and insight from other people … • **Several** sources often prove more valuable than one.	• I'm very vulnerable when I act on this belief … • Yet, my ego may whisper "go ahead."	• I'm also vulnerable in this area … • I can expect surprises "out of the blue."

PEOPLE WHO ASK LOTS OF QUESTIONS AND THEN REALLY LISTEN TO OTHERS ARE THE "HEROES" HERE. THEY GET TWO THINGS: ACTIONABLE INFORMATION *AND* THE RESPECT OF OTHER PEOPLE.

B. Focus on What You Can Influence

Both Arthur Ashe and Stephen Covey made the practical suggestion that we—whether groups or individuals—focus our energies and attention on those things we can **change** or, at least, **influence** (*Figure 12.1*).

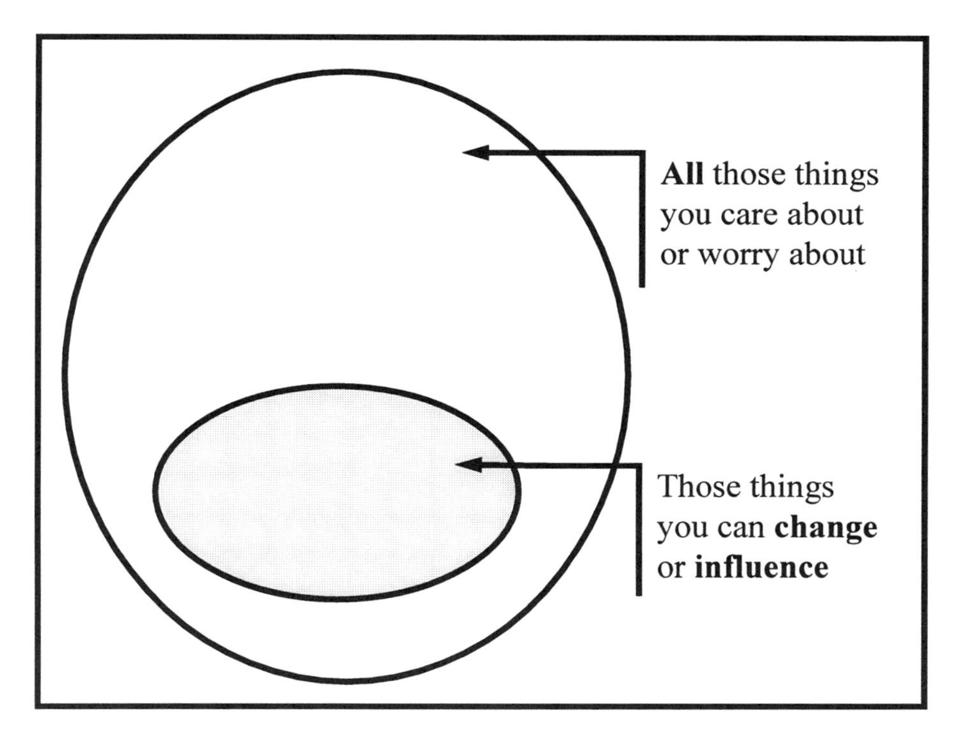

Figure 12.1: What You Can Influence

To spend much time and energy on those things you **cannot** influence, generally leads to frustration, wasted effort and perhaps a pessimistic view of life.

As you focus, **successfully**, on those things you **can** influence, your sphere of influence is likely to grow larger.

> *Be the change you are trying to create.*
>
> ... Ghandi

C. To Change Others, Behave Differently

Biologist Van Bertelanfy gave us insight into **systems**, based on his studies of systems in nature (as discussed by Watzlawick, Weakland and Fisch in their book *Change*, WW Norton & Co., New York, NY, 1988). One very practical aspect of this effort tells us that when one part of a system changes, other parts will also change (*Figure 12.2*—adapted from ***Change***).

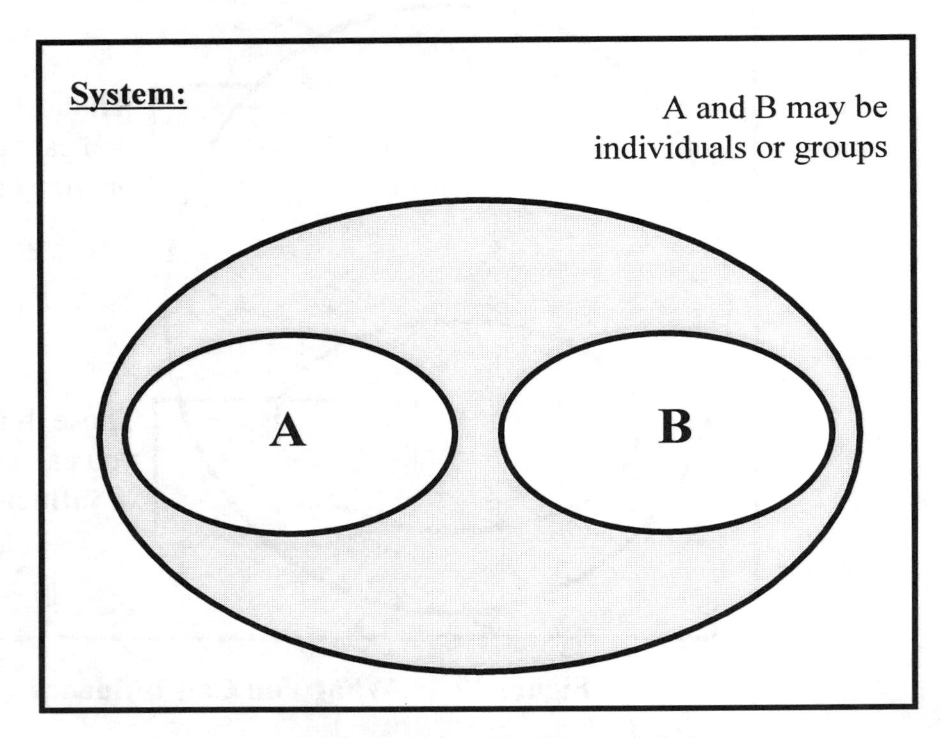

Figure 12.2: How To Change Others

If A and B have some sort of relationship, you may say they **are** a system or are **in** a system.

While A cannot (usually) change B directly ...

- A can change A's own behavior in ways B will notice.
- B will usually, and eventually, change his/her behavior in response to A's change.

Chapter 13

SPIRITUAL PATH, PRACTICE OR DISCIPLINE

Personal development is more dependent upon one's **spiritual** path, practice or discipline than many of us realize. And more people are noticing the importance of spirit than ever before. For example:

- Gallup polls show that a larger percentage of the American people attend church or temple today than in 1940![1]
- An **Atlantic Monthly** study and report indicates such attendance today is more often motivated by personal commitment than acts of conformity.[1]
- Numerous studies have shown that spiritual faith and practice have positive impacts on individuals, families and society. Regular church, synagogue, or temple attendees ...
  - are 50% less likely to suffer mental problems
  - have 50% fewer divorces
  - are 71% less likely to become alcoholics.[1]
- Though seldom on the front page, newspaper columnists regularly report the proven power of prayer on behalf of others to improve their chances of healing and recovery from serious illness, accidents, surgery and medical treatments (in comparison tests).

Surgeons and physicians have reported and/or written books on the **healing power** of faith, spirituality, support groups, prayer and of one's own focused, positive mental attitudes.

---

[1] ***Is Progress Speeding Up?*** by John Marks Templeton; Templeton Foundation Press; 1997.

Recently, the magazine ***Fast Company***[2] presented the work of Dr. Ben Carson, one of the world's most celebrated brain surgeons, and his techniques for coping with pressure, planning for problems and dealing with risk—very simply, prayer, faith and preparation. He says, "you can understand why I'm a believer. I have seen miracles." He shares some of this in his book: ***Think Big*** (Harper Paperbacks, 1992).

Touchstones for Faith

Only a hermit could avoid noticing that "progress" in most areas of our lives is at best a **mixed blessing**. Along with advances in technology, art, communication, medicine, work savers and productivity, we also face increasing crime, drug use, divorce, sexual promiscuity, gambling, etc., especially among the young.

For example: While the U.S. "health care" system leads the world in research, new drugs, hospitals, and expenditures, we rank only **40th** in the world for a healthy citizenry![3] (The problem has more to do with our lifestyle.)

In the face of so much bad news, I find it necessary to look for some indications that we as a society have not been abandoned by the Source, Yahweh, the Lord of Life, Great Spirit, Allah or the "Hound of Heaven."[4] And when I look, I do find some indicators. Here are some that work for me:

- For those who work in science or technology, note the faith of a Madam Curie or an Albert Einstein. Einstein discovered relativity and changed the "laws" of physics because he said he **"knew"** a caring creator would link time and space.
- At a time when Americans have a greater percentage of our population in prisons than ever before, we also see the blossoming of faith and new, creative ministries in and from **prison** walls as well as for crime victims' and prisoners' families.
- Even with all the downsizing, outsourcing and other turmoil in business and industry, spirituality is breaking out in practical ways at **work** through phenomena like "Servant Leadership" inspired by Robert Greenleaf of AT&T.[5]

---

[2] **"This is Brain Surgery."** ***Fast Company***, February-March 1998 Issue, pages 147-150.

[3] Data from the World Health Organization.

[4] From the well known poem by Francis Thompson.

[5] ***Servant Leadership*** by Robert K. Greenleaf, published by Paulist Press, Mahwah, NJ, 1977.

- In some former communist nations of Eastern Europe and Asia, **open** religious worship is thriving again. And, in other countries, religion is being intensely persecuted but is thriving anyway.

To my way of thinking, the Spirit is up to the same old "tricks" as always: speaking to humankind in **unexpected** ways as well as to and through **unlikely** people such as … Mother Teresa, Thich Nhat Hanh (Zen monk and reconciliation activist), and Sir John Templeton.

Spiritual Quest

Peter Vaill, management consultant and teacher, wrote the book ***Managing as a Performing Art*** in 1989 (Jossey-Bass). For a **business** book, his chapter on spirituality and leadership was courageous, unapologetic and unusual in 1989 and **still is today**.

He believes that **true** leadership is indeed **spiritual** leadership because it is primarily concerned with bringing out the best in people. Here are some other personal, practical, helpful points he makes there:

- Each of us needs to think of ourselves in **spiritual** terms as well as the psychological and economic terms the 20th century has taught us (perhaps too well).
- Spiritual renewal in our culture and organizations is **everyone's** problem (and opportunity). And it must begin with each of us, wherever we are on the spiritual spectrum.
- "It is indeed a **quest**. I will discover spirit in myself and others as I search." Living, he believes, is all about the search for spirit.
- Vaill notes that this search is not about bringing spiritual "answers" to others because they are on their own searches. Yet, as fellow searchers, I think we can and do **help** one another.
- He paraphrases George Washington, who believed that as part of our search, we may find "the same spirit does animate the whole (of humankind)."

 He believes that religious faith won't solve all or any of our problems; rather it helps us to deal with them in a world of changing, but never-ending, troubles.

Spirit at Work

Firms, small and large, from Schneider Engineering (Indianapolis, IN) to giant Federal Express, are putting to use the practical principles of **Servant Leadership**. Greenleaf believed that true leadership begins with the **desire to serve**, rather than to control, tell, direct, manage and so on.

Herb Kellaher, President of Southwest Airlines, apparently also puts those principles to work if we can believe the business press … and I do. His example of caring leadership ricochets through the Southwest organization, down to the pilots, agents and attendants who treat Southwest passengers well. No wonder they set so many airline service records.

We need to notice that the lessons here are passed on through **deeds** more than words. For me, one of the differences between spirituality and religiousity is this focus on service, on walking the talk, on meeting needs, rather than preachments and judgments. I suppose we need both, but I believe we need more of the former. We need to try to "walk" our spiritual beliefs, as well as talk about them.

Walking Your Spiritual Path

Our colleague and friend, Isabelle Healy, has a consulting enterprise in Cincinnati called **Working Together** (513/351-5140; ihealy@eos.net). In Isabelle's ***I Was Thinking …*** newsletter (No.11, Spring, 1999), she shares insights about "Walking Your Spiritual Path at Work." She's given me permission to reprint some of those thoughts here …

"Our work is spiritually significant. We all have gifts that we have received for our own enrichment and to contribute to a world that is bigger than ourselves. How we contribute our gifts is our work.

"We all have a work. 'Work' is not the same as 'job.' Our work is what we contribute spiritually outside ourselves or beyond ourselves. Retired people who consciously choose to use their time and energy for spiritual pursuits have a work. Stay-at-home moms who approach the care of their children and their households in a spiritual way have a work. And those who have a job can make this their 'work,' their particular way of contributing to the presence of Spirit.

"We have something spiritual to bring to our work only if we have an inner, spiritual life. Even though our external work is spiritually significant, there is something that is more significant: our inner life. This is a hard lesson for us Americans who have been taught that doing is the measure of our worth and doing that makes money is even more worthwhile.

"One of the most time-honored ways to nurture our inner life is silence. The wise ones from all spiritual traditions have urged those who want to develop their spiritual nature to practice silence. Every day, have some time in your life when you turn off all the external sounds—the radio, TV, even voices of your spouse, your kids and the other people in your life—everything. You might sit or walk, but don't do anything that would occupy your mind. Then see what happens."

Thankfulness for Life

One of the things I've begun to notice is that individual happiness **isn't** a function of what people have or lack. It's more of an inner happiness (if it's there) and/or an appreciation for the many wonders of life.

Occasionally, I'm in the company of a world-class whiner ... actually, I used to be one myself. You've heard the expression "rose-colored glasses." Actually, I think a bigger problem is a person who seems to look at his/her world through "poop-colored glasses."

Wish I could remember the writer who suggested that we each write our own "gospel." By that he meant a list of things and experiences we have had, or people we have known in life, for which we are thankful. I did. It was one of the most profound experiences of my life. I filled five pages, even though my writing was concise!

Do yourself a favor and give it a try. Start here (but you'll need more paper) ...

____

____

____

____

____

____

____

The next time you feel that life is mean or completely evil and that there is no good in it for you or anyone else, try this: make a list of some of the beautiful things you have seen, the breathlessly kind things people have done for you without obligation, the gracious moments that have turned up in the week's encounters. Memory is one of God's great gifts to the human spirit without which neither life nor experience could have any meaning.

... Howard Thurman
African-American theologian and writer, 1899-1981

Intriguing Thought

The Native American cultures of the Southwest, such as Northern New Mexico's Tewa and Tiwa Pueblos, have survived almost 400 years of invasions beginning with the Spanish.

For these people, "there is no specific word for religion; it is simply the proper way to live one's life, a life devoted to the well-being of the tribe and the family."[6] Their dances are spiritual ceremonies, usually expressing their desires for "survival and for the **spiritual well-being of the Earth**"[6] on which we all depend and which we need to care for better than we often do.

Why the Leaf?

You may have wondered why a **leaf** is the icon for this chapter. It's been our business logo for 25 years, developed for us by artist Joe Pagliaro. Joe and Carol have long been spirit-filled friends.

The leaf logo reminds us of our love for nature in most of its forms, especially **trees**. Trees are models of beauty and service in so many ways, from providing shade and lumber to sustaining birds and fresh air. Some native Americans regard trees as the "standing people" ... a nice metaphor. And, as I said, trees are good models of service for **us** ... the "moving people!"

Good to Know

In my practice and in my community, there are several first-generation Americans who follow the Sikh religion, begun in the Punjab region of India and Pakistan in the 15th century. A minority that is sometimes persecuted for their faith, the Sikh religion was founded by Guru Nanak.

Sikh beliefs include these:

- There is one God.
- In service to humanity is "found a seat in the court of the Lord."
- The equality of all men and women, opposing class systems.
- Sanctuary must be provided for those in need or danger.[7]

---

[6] ***Eight Northern Indian Pueblos Visitors Guide - 1998***

[7] ***The Cincinnati Enquirer***, Section D Tempo, November 12, 1998.

Religious Tolerance

Dr. Paul Marshall, senior fellow at Freedom House's Center for Religious Freedom in Washington, DC, has studied religious intolerance in depth. **He urges us to make religious rights prominent among human rights!**

Betty Eadie, author of ***Embraced by the Light*** (Gold Leaf Press, Placerville, CA, 1992), is—as I am—a follower of Christ. She tells us that we need different churches and traditions to accommodate the wide diversity of people, their needs and interests and stages of development.

Your Thoughts

What are some of **your** thoughts about spirit or **your** spiritual path, practice or discipline (if you have one)? ______________________________

____

____

____

____

____

____

____

____

____

____

____

____

____

____

> *Bidden or not, God* ***is*** *present.*
>
> ... Carl Jung

Chapter 14

SETTING WORK, CAREER AND PERSONAL GOALS

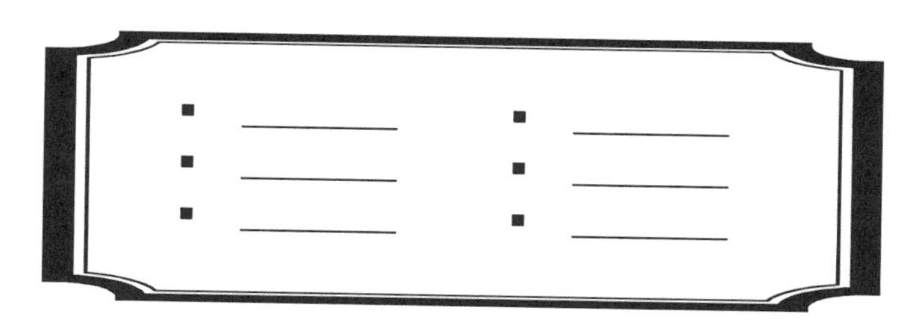

Over many years of study, consultants and researchers have discovered that relatively few people (3% to 5%) have explicit and written goals.

> **Those who do have explicit goals are far more successful in life than most of us! However you define "success"—financial, familial, technological, entrepreneurial, or contributing to others—they are clearly more successful.**

Perhaps the most important things to be said about setting goals for work, career and personal success are these:

- **Having clear goals can make a difference to you; that is, goals can really change things.**
- **Goals serve you best when they are in alignment with your life mission or purpose (Chapter 3).**

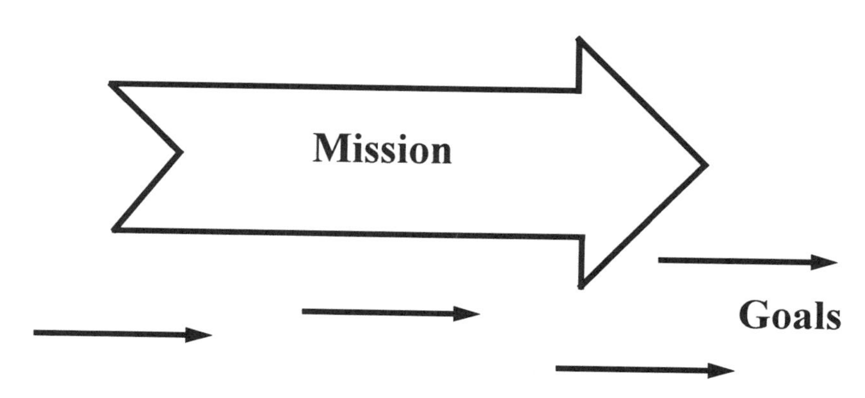

Speaking for myself, I've generally done better at **having** some goals than at having my goals be in **alignment** with my purpose. A couple of examples may help you see what I mean:

- My children were (and are) very important to me. In my efforts to "get everything done" (career, continuing education, home chores, etc.), I didn't spend as much time with them as I wanted to, especially when they were very young.

- Pursuing more education while working seemed like a good investment. I didn't give enough thought to which graduate studies could be most helpful in the future, and for **me**, an MBA would have been more useful than the MS I pursued.

To say it another way, I've met most of my goals, but some of them weren't in alignment with my mission. They didn't serve me, my family or my career as well as some **other** goals would have! So, check **your** goals for alignment with **your** mission.

Opportunities for Goals

Or, if you decided to write some goals, what might they be about? Here are some possibly fruitful places to look for **potential goals**—a menu of goal opportunities:

- An aspect of your mission
- A challenging project at work
- A hoped-for promotion or job change
- A family-related need or hope
- An important, needed skill set
- A health-related need or hope
- A major threat or difficulty
- A spiritual need or spiritual growth
- An avocation, hobby or interest

Figure 14.1: Goal Opportunities Menu

Helpful "Feedback"

A particularly good source of possibilities for work and career goals is **helpful feedback**. Forgetting all about the business jargon, "feedback" simply means information on **how we're doing**. "Helpful" feedback means:

- **Behavior-based:** Focusing on your observable behaviors and actions rather than your intentions or attitudes.
- **Specific:** Rich enough in descriptive detail that you can "see" it in your own mind's eye.
- **Actionable:** Something you can actually change (if you want to).
- **Objective:** Information that isn't tainted by anger or slanted to patronize you.

This kind of feedback is hard to get because people (including bosses) often don't know how to give useful feedback and often hate to give feedback at all. Here are some tips for getting "helpful feedback" for yourself:

- **Ask** for feedback. It will usually help you get the kind of feedback that's worth considering for your goals.
- Ask **several** people ... boss, colleagues, customers. This helps you get more objective and credible feedback.
- Ask for **specific** feedback, such as:
  - "What new skill(s) do I need to develop?"
  - "What are your concerns on Project 107?"

Forming Your Goals

Contributors to NLP offer some questions to help us develop well-formed goals. These questions require us to think in terms of **desired outcomes**:

- What, exactly, do you want as an **end result**?
- If this goal is triggered by a problem with others, what do you want **for yourself**?
- How will you **know** if you get it?
- How **soon** or how **often** do you want it?
- **Where**, or with **whom**, do you want it?
- Will this outcome require some **costs** or tradeoffs?
- Will this goal **conflict** with another goal?
- If so, how will you prioritize or resolve the issue?

Test kit for well-formed goals:

- Clear outcome?
- How you'll know?
- Time frame?
- Where and with whom?
- Tradeoffs?
- Priority?

Sample Goals

- **Project 107:** On time, under budget, no claims, no lost-time accidents.
- **Behavioral:** "My conflict management skills are developing very well" (see *Table 10.1*).
- **Significant** new responsibilities through a promotion here or a new role elsewhere by June 1, 2001.
- **Weigh-in** at 185 pounds or less through exercise and nutrition by January 1, 2000.
- **At least 1.5 hours** per week of quality time with Sarah and Tom.
- **Participate** in at least two professional development workshops, conferences or seminars every year.

Other Thoughts

Here are some other potentially helpful things to consider about your goals, goal-setting, and accomplishing or achieving your goals:

- It's important to keep our goals **clearly in mind** by reviewing them often. So, put them in several places where you'll be reminded of them.
- For **behavioral and skill** type goals, consider the process suggested in Chapter 10, *Table 10.1*, Phase 3. This process draws on the powerful resources of the sub-conscious mind (our mental "software").

Your Own Goals

Try your hand at some goals for yourself now. Consider areas like mission, career, job, challenges, family, health, skills, faith, and avocation(s).

____

____

____

____

____

____

Your Own Goals *... continued ...*

Chapter 15

THINKING ABOUT A JOB CHANGE

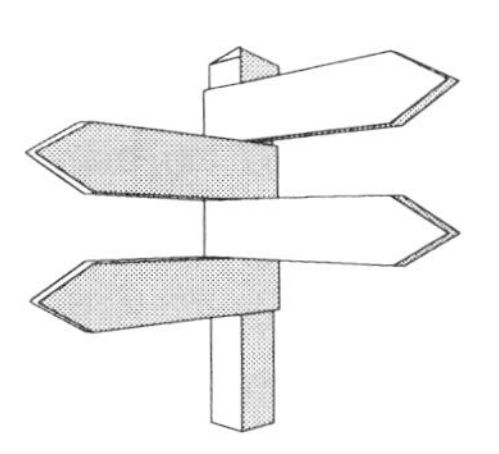

One of the most surprising aspects of my consulting practice has been helping people consider a **job change**. It's surprising to me because that's not something we've offered to clients. Yet, often clients have asked for our help in thinking through job opportunities or dilemmas with their current jobs.

Because of that, and driven by my curiosity about careers, I've come across a number of approaches to helping people think about their lives and livelihoods, their interests and their careers.

In this chapter, you'll find several approaches that have been some help to our friends and colleagues. They may also be helpful to you at some point. But, before we consider those, let's review some previous chapters relevant to this issue!

The following chapters and topics may well be relevant prior to considering a job change:

Chapter 2: *Stages of Human Development*

Chapter 3: *Your Life Goal, Aim or Purpose*

Chapter 7: *Appreciating Differences in People*

Chapter 10: *Dealing Well with Adversity*

Chapter 14: *Setting Work, Career and Personal Goals*

Your Support System

Each of us needs a strong **"support system"** almost **every day**, whether we know it or not, or acknowledge it or not. We need our grocer, our utility provider, our employer(s), our pharmacist and health professional among many others.

Focusing on our world of **work**, we need a support system there as well. At times when we're considering a job change, our support system becomes even more **critical**, as we reflect on our hopes, needs, skills and options.

Nancy J. Miller, both a client and colleague at different times, introduced me to the following notion of **work-related support systems** ...

Table 15.1: Work-related Support Systems

THE KIND OF SUPPORT YOU NEED:	WHO ALL HELPS YOU IN THIS AREA OF SUPPORT:
1. **Friendship** and personal companionship	
2. **Listening** and providing a "sounding board"	
3. **Professional** peers who have similar work interests	
4. **Wisdom** and insight into job-related situations (mentoring)	
5. **Problem-solving** assistance, both career and personal	

Table 15.1: Work-related Support Systems *... continued ...*

THE KIND OF SUPPORT YOU NEED:	WHO ALL HELPS YOU IN THIS AREA OF SUPPORT:
6. **Challenges** to your assumptions, thinking and conclusions	
7. **Connections** to resources or people who can get them	
8. **Networking**; linking to possible future employment opportunities	

Turn-ons and Turn-offs

Stan Hinckley, Rob Daly and Charles Smith did a lot of creative human resource development work for Procter & Gamble in the '70s. I believe this notion of identifying "turn-ons" and "turn-offs" came from their creativity.

Draw two lines down a page to create three columns, like so ...

A	B	C

➚ **Step one:** In Column A, identify all the **roles** you fulfill in your current life or recent past, such as:

- nurse, engineer, salesperson
- parent, uncle, grandparent
- committee chair or member
- little league coach or referee
- manager, group leader, or boss
- volunteer fund drive solicitor
- homemaker, part or full-time
- musician, artist, craftsperson.

➚ **Step two:** In Column B, identify **what you're doing** when you feel most turned-on and alive.

➚ **Step three:** In Column C, identify **what you're doing** when you feel turned-off, bored, hassled, etc.

➚ **Step four:** Look for similarities and/or patterns in Columns B or C.

You may find some helpful clues about **what kinds of activities** turn you "on" or "off." If you do, use these clues to help evaluate job prospects, including your present job!

Life Purpose (... again)

Whenever friends or clients talk about their job satisfaction (or lack of it), I sometimes sense that they're doing work they love so much they'd even do it for less money. Other times, my sense is that their work barely satisfies their need for a wage or salary!

Those in the latter group would surely benefit from working to define, clarify or revisit their **life purpose**, aim or mission, as discussed in Chapter 3, and then to use that "mission" to take the measure of prospective jobs or their current job.

> **Suggestion:** Try to stay open to the possibility that your current job may **become** more what you want if you can make some minor changes!

Career Anchors

Edgar Schein, in one of his excellent contributions to the Organization Development Series of Addison-Wesley, gives us a helpful view in ***Career Dynamics: Matching Individual and Organizational Needs*** (Addison Wesley, Reading, MA, 1978).

He identifies **five different "career anchors"** that may draw individuals to a satisfying kind of career:

- **Technical or Functional Competence**
  the desire to be **expert** in a particular technology or work function such as finance or marketing.
- **Managerial Competence**
  the desire to lead, manage, direct, or guide others in an enterprise, an organization or a part of one.
- **Security**
  the desire to have work that we know (or believe) will always provide a sense of predictable security.
- **Creativity**
  the desire to work in an arena or setting where one's creativity can be fully exercised (and pay the bills).
- **Autonomy and Freedom**
  the desire to "be one's own boss"—to decide **what** to work on as well as how, when and where.

When I've worked with a client or/and friend using these five "career anchors," they have generally been able to rank them for themselves rather easily.

Using that ranking, they then have another tool to evaluate their current position as well as any other job prospects or opportunities.

Skills Inventory

When you find yourself in certain situations like these ...

- considering a promotion or transfer
- evaluating a new job offer
- assessing current job satisfaction
- finding oneself unemployed

… you could benefit from doing what is called a "skills inventory." Some of the following ideas come from Crystal and Bolles in ***Where Do I Go From Here with My Life*** (Ten Speed Press, Berkeley, CA, 1983).

- **Step one:** **List** your accomplishments or achievements, particularly those that give you a lot of personal satisfaction.

- **Step two:** **Identify** the skills and abilities those accomplishments required, such as:
  - relating well to others
  - analyzing complex problems
  - organizing work and people
  - working long hours
  - stretching a dollar
  - balancing many tasks
  - writing reports well, etc.

- **Step three:** Rank those skills from top to bottom in terms of how **competent** you are.

- **Step four:** Rank them again, but this time in terms of how much you **enjoy** using them. (These can be called "motivated skills.")

Now you have another tool by which to measure potential careers or positions, as well as the one you're in now.

Your Signature Path

Your Signature Path is an incredibly helpful workbook and guide by Geoff Bellman (Berrett-Koehler Publishers, San Francisco, CA, 1996).

It's written, says Geoff, "for people who reflect and act upon what their lives are about, who intend to have a hand in their own destinies, and who aspire to become their better selves."

And, indeed it is. It's fun to read and produces helpful insights of the kind one needs when thinking of a career or job change, whether early or late in life. Having seen a lot of these kinds of books, I was quite surprised at how helpful Geoff's book was to me! I highly recommend it.

Chapter 16

HELP FOR MANAGING TIME (AND LIFE)

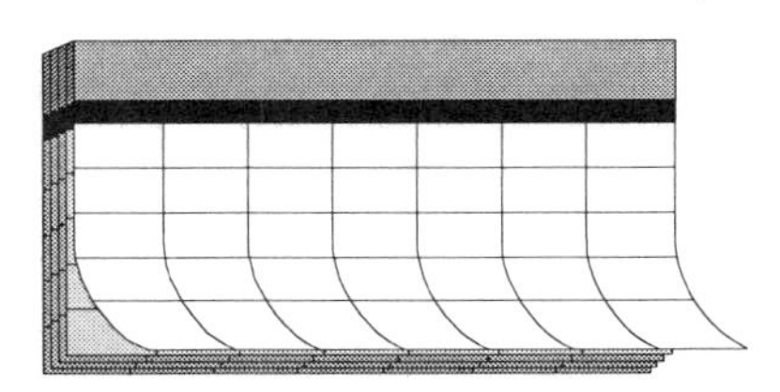

Over many years, I've come across some nifty "time management" concepts, tools and tips in more than a dozen good books and hundreds of articles. They were all interesting, some were helpful, yet most of them missed some of the most **fundamental notions** of time management. But first, a story:

> About 1980, I became particularly committed to a conference center that was supported by two church denominations. After a long struggle, this center hired a great manager in Fran Brown, who brought in new guests, new services, and increased revenue as well.
>
> As volunteer in charge of "grounds," I spent 10 years planting, painting and installing facilities like volleyball courts, baseball backstops and creek bridges. Since I was using a lot of my own labor and money, you can bet I used every time management tip I had!
>
> Then one of the denominations recalled their loan and the volunteer board decided they must sell the conference center. All our work was leveled and planted in grass! This wasn't my only hard lesson in time management, but it gave me a whole new perspective.

I'd already learned about the importance of **balancing priorities, leverage and timing,** as well as **"to do" lists and planners**. Then the conference center experience taught me the importance of **purpose and focus**.

Purpose and Focus

So much of the time management lore deals with setting priorities, deciding what's urgent, important, and so on. This is all true, but "urgent and important" to whom? And why? We need a way of **knowing how to decide** about urgency and importance.

Chapter 3 helps to clarify one's life **purpose, aim or mission**—or at least some ways to begin that process. That clarification may be fast for you but, more likely, it will evolve over the years.

Whatever **clarity of purpose** you may have at any point in your life can help you. It can serve to guide you in the same way a compass can help you to orient a map. **Purpose** may help you sharpen the focus you need, showing you what things are **in focus** and which are **not**.

> There is almost always more to do than there is time to do it. Maybe 90% of time management has to do with clarifying what to **do** and what to
>
> - skip over altogether
> - leave until another "time"
> - give others an opportunity to do.
>
> And, there is no better gauge for this than your aim or purpose in life, even if it's still a bit murky at this moment.

I've been working, periodically, to clarify my own purpose for about **40 years now**. Here's how it seems to me at this moment:

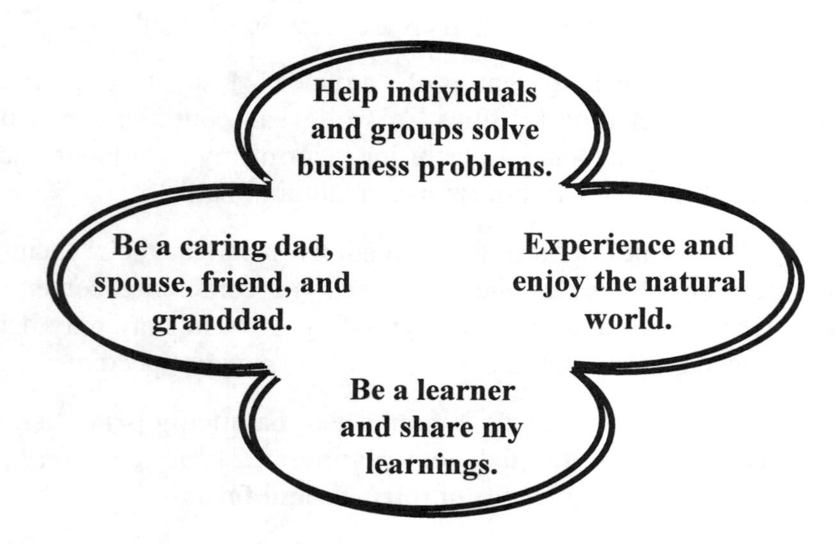

Figure 16.1: Mel's Purpose

And so, being the groundskeeper and facilities fixer for a conference center probably didn't fit my purpose very well!

Reflection on Lifestyle

This chapter has been a vacation enterprise while Carol and I have been touring the Amish region of Northeastern Ohio between Zoar and Millersburg. And in between, there is Charm, Walnut Creek, Sugar Creek, Winesburg and Dover, plus many country roads and farms.

We enjoy the Amish folkways: people traveling on bicycles and in horse drawn buggies, wearing straw hats or starched bonnets, working their farms by hand and horse power, running tasty restaurants, and selling well-made crafts and goods.

How do they "have time" to live this way?

Well, I'm certainly no expert on the Amish, but any careful observer can see they take time to visit and to enjoy their homes, shops, farms and neighbors.

I recently read that when their children are young adults (late teens) they are encouraged to run with others their age, free of family and rules, so they can decide for themselves if they want to follow the "old ways."

My guess is they know what's really important—family, work, faith and neighbors. And "everything else" isn't a priority with them.

Simplify, Simplify

Simplicity expert Elaine St. James' interview in the June-July 1998 issue of ***Fast Company*** magazine is good medicine for your time management **and** your peace of mind.

Here's a sampler from St. James:

- No one can maintain more than **three** priorities. Your job or career is one. If you have a family, that's two. Which leaves one more.
- Possessions are nine-tenths of the problem. **Stuff** doesn't just cost money, it also takes time, which is what people say they **really** want.
- Figuring out what **really** matters to you (sounds a little like "purpose" again) will help you make time for that.
- Then, learn to say **"no"** to people and to expectations that don't fit, don't help, or don't feel right.

Personally, I'm getting better at remembering the advice a colleague keeps pinned on her refrigerator:

Remember: Things are easier to get into than to get out of!

Balancing Priorities

Several folks have offered tips for this, and my own personal favorite is Alan Lakein's **"urgency" vs. "importance" matrix** (***How to Get Control of Your Time and Your Life***, published by Signet, a division of Penguin Books, New York, NY, 1974). Here's my version of it:

IMPORTANT AND URGENT	IMPORTANT BUT NOT URGENT
URGENT BUT NOT IMPORTANT	OTHER STUFF

Figure 16.2: The "Urgency" vs. "Importance" Matrix

The tyranny of "urgency" will usually assure that the urgent things get done somehow, even if a tad late!

Of more concern are the **important but not urgent things**, such as:

- coming up with the "next" product or service.
- developing our leadership "benchstrength."
- building our retirement "nest egg."

(See the section on "Reversing Procrastination" later in this chapter.)

Of some interest is the "other stuff"—neither urgent nor important. Why would you do those at all? Well, usually because they are fun, relaxing, a change of pace, a stress reliever and just enjoyable. These activities can enhance our lives, **unless they are stalls** to avoid the urgent/important stuff.

Leverage and Timing

Many of us have had the experience of trying to do a challenging task and having all the "bad breaks." Other times it seems the breaks go **our** way and the job goes smoothly. Some part of this is just plain luck, good and bad.

But, some of it is **optimum timing**, some of it is **leverage**, and some of it is **critical mass**.

The simplest way to allow for better timing and leverage is to have a wider, **multi-tasking approach** vs. a sequential-tasking approach to moving projects along toward completion.

By "multi-tasking," I simply mean setting up many tasks so you can do them when the timing is to your best advantage. This approach requires patience, but it has rewards that make it worthwhile, including:

- spontaneity and variety
- paths of least resistance
- less stress and frustration
- getting more done **overall**.

The multi-tasking approach is made easier by good "list-making" which is covered next!

To-Do Lists

"Lists" can become part of the problem if you let them get out of hand. For example, we see:

- people who get "lost" in their (too) many lists.
- big personal planning notebooks that require lots of writing.
- planning software that consumes hours, then isn't where you need it.
- post-it notes stuck to anything that doesn't move, then they fall off.

In fact, systems like these can and do work very well if you fine-tune them to **your work** and **your personality**.

Here's my personal favorite "system" which has evolved over time by freely borrowing good ideas from clients and colleagues:

- **Pocket-size, spiral-bound calendar**, good for a year, with about an inch for each day. In it I write both professional and personal upcoming events (in pencil, easy to change). Available in office supply stores.

- **3" x 5" plain file cards** for jotting down tasks, calls, errands, etc., that don't relate to a particular date, they just need to be done. Some are project-specific, some are personal. I toss them when the list is "done."

Both of these fit easily in my shirt pocket and pants pocket or briefcase and wallet. They are always easy to take with me to the office, on the road, and even on vacation, as well as at home.

Reversing Procrastination

An interesting difference between those who procrastinate a lot vs. a little is how they "see" and feel about a to-do task.

- "Procrastinators" see a to-do task as large, looming, never-ending, and difficult, and are feeling weary before they even start.
- "Proactives" see a to-do task as **already** done, and are starting to feel good about that even as they begin the work.

Most of us procrastinate about **some** things, and we can and do enjoy doing **other** things. Changing ourselves toward being more proactive more often can be as simple as changing how we "see" and feel about a task we have ahead of us.

More Help

Time Management for Engineers and Constructors, 2nd Edition, 1998, by Ray Helmer is a good resource available from ASCE Press.

Getting Started

1. Jot down your life purpose or mission as it stands now. Can it be **helpful to you in focusing or setting priorities**? ______________________

   ____

   ____

   ____

   ____

   ____

   ____

   ____

   ____

   ____

   ____

   ____

2. What on-going activities and/or "stuff" do you need to get **out of your focus and your life**? ______________________________

3. How can you improve your **to-do listing and follow-up "system"** … either starting one or simplifying what you have? ______________________________

4. What really important to-dos do you currently procrastinate about and need to begin, **now**, to **visualize differently and feel better about**? ______________________________

> *There is more to life than increasing its speed.*
>
> … Mahatma Ghandi

Chapter 17

CREATIVITY AND RESOURCEFUL THINKING

Creative thinking is a **much more important skill** than most of us realize—both in our personal lives and at work.

Generally speaking, North Americans are great problem-solvers and creative thinkers. But I believe this advantage is mostly a result of our **diversity** and our relative **freedom** to do as we choose. Both are spurs to creativity.

So, beyond that, what might we do to enhance and spur our own creativity? (**Good** references on creativity and resourceful thinking are hard to find.) Here are several that you may find enjoyable to read and use:

- ***Using Your Brain - For a Change*** by Richard Bandler, published by Real People Press, Moab, UT, 1985.
- ***The Creative Spirit*** by Coleman, Kaufman and Ray, published by The Penguin Group, New York, NY, 1992.
- ***Change*** by Watzlawick, Weakland and Fisch, published by W.W. Norton Co., New York, NY, 1988.
- ***Brain Power*** by Karl Albrecht, published by Prentice Hall Press, Edgewood Cliffs, NJ, 1980.

Creativity Clues

From these favorites, plus many other sources, I've collected a practical set of tips for almost instantly improving my own resourceful and creative thinking. Perhaps several will also be helpful for you:

[1] **Get into a resourceful state!** What I mean by this includes: **not** being stressed out, **not** sitting back in your recliner, **thinking positively** about the problem or situation, **remembering a time** when you were creative! Anything that enables you to feel resourceful will help.

[2] **Study up on the problem or need!** Study up, but not necessarily books. Anything that helps you learn about it will help. For example, on a home repair problem, I like to wander through a hardware or building supply store! For a client's challenging strategy problem, I "wander" through my strategy files!

[3] **Ask (lots of) other people!** My model for this is Jim Lockwood. Jim is an accomplished executive and very creative engineer. But I never met Jim when he didn't have questions for me (or anyone). And he **listens**! Too many of us fail to use the resourceful people around us because we don't want "advice."

[4] **Leave the problem "open" for a time!** Try not to rush to an "answer" just so you can get comfortable with it. This wait allows time for your subconscious mind to work, for others to contribute, for external stimuli to help, and so on! (This is **not** procrastination if you actually think about it.)

[5] **Focus a brainstorming session on it!** If you have time, and a group is available, and assuming the problem or need is worth it, **groups are great!** Just before launching our consulting enterprise, Carol and I were at a social gathering of close friends. Their ideas for our office-to-be were truly helpful.

The STOP Process

Whether you're working alone, with one colleague, or as part of a task team or group, the "STOP" Process will help you get out of the circular, non-creative thinking rut.

STOP is a little different than the "POP" Process from my book on teams (***Collective Excellence***, ASCE, 1992). The STOP Process is shown, very briefly, in *Figure 17.1*.

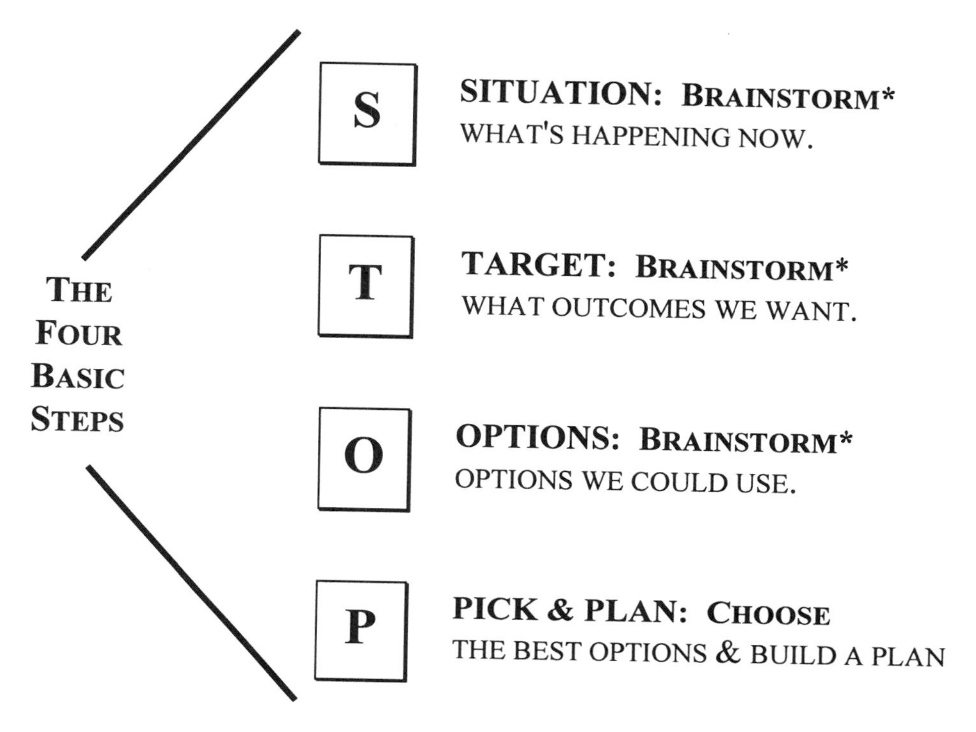

***Remember:** Brainstorming takes **all** contributions with**out** criticism.

Figure 17.1: The STOP Process

Other Creativity Enhancers

Like so many other things in life, thinking resourcefully and creatively is mostly a matter of …

- **really wanting to** (desire)
- **believing you can** (faith)
- **getting started on it** (action)

… much more so than "techniques."

That said, there are some approaches and techniques that can enhance your abilities in the area of creative and resourceful thinking. Some I've found personally very helpful are shown in *Table 17.1* following.

Table 17.1: Creativity Enhancers

INSPIRED BY MIKE VANCE (FORMERLY OF DISNEY)	INSPIRED BY DOUG HALL (FORMERLY OF P&G)	INSPIRED BY TONY BARZUN ON "MIND MAPS"	INSPIRED BY ALEX OSBORNE ON "BRAINSTORMING"
Brainstorming often benefits from a bit of structure like this: • Identify topic "clusters." • Add items to each topic cluster. • Add new topic clusters as they come up.	Real creativity requires long hours and strong effort. It requires stimulation "outside the box." You must steep yourself in the topic beforehand. Keep at it until you produce clearly creative results!	The mind isn't usually linear in its creative mode! So, utilize nonlinear thinking. Go in all directions. Linkages are often illogical. That's OK and even good!	Use mental gymnastics to get "out of the box," such as … • turn it inside out • upside down • go inside it • shrink it • explode it • relocate it • ask what if? • and so on.

Next Steps

Ever so quickly, **jot down some things** you might easily do to begin to actually enhance your own personal creativity and resourcefulness!

- Think of **several times** in your life when you experienced being resourceful and creative …

<u>**Situation/Experience**</u>	<u>**How I Felt**</u>
______________________	______________________
______________________	______________________
______________________	______________________
______________________	______________________
______________________	______________________
______________________	______________________

- Think of several **current challenges** where you'd be glad to be creative and resourceful …

1. ____

   ____

2. ____

   ____

3. ____

   ____

- Where could you **"wander" (actually or figuratively)** to gain ideas and/or insight on these challenges?

1. ____

   ____

2. ____

   ____

3. ____

   ____

Who could you **chat with** about these same challenges to gain new perspectives, ideas and/or insight?

1. ______________________________

____

2. ______________________________

____

3. ______________________________

____

Any **other** creative thoughts about these challenges?

1. ______________________________

____

2. ______________________________

____

3. ______________________________

____

Chapter 18

INFORMATION AND DECISION MAKING

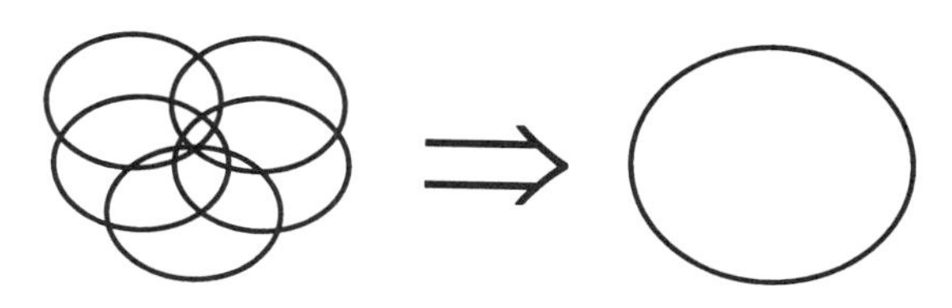

> *Never before has so much technology and information been available to mankind. Never before has mankind been so utterly confused.*
>
> —KPMG ad in ***Fast Company*** Magazine

The following notion comes from John Nelson, an investment manager for several groups and board member for one of our clients …

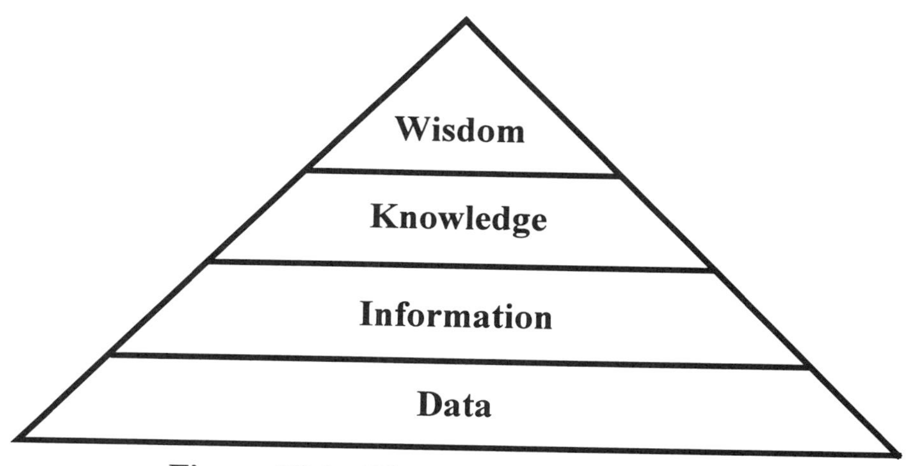

Figure 18.1: Hierarchy of Information

Risk: Probability of deviation from what is expected.

⇒ Data → Information → judgment → risk → reward.

⇒ **Good** information → good judgment → lowers risk & increases rewards.

⇒ **Poor** information → poor judgment → increases risk & lowers rewards.

Data Traps and Tricks

Many writers and researchers (Deming, Tracy, Fisher, DeBono, others) have contributed to my understanding of some pitfalls in the area of "data":

Table 18.1: Data Traps and Tricks

We tend to overvalue whatever data we **have** vs. data we **don't** have.	Our mental maps or paradigms cause us to **"see"** some data and **"not see"** other actual data.
We tend to overvalue data that's important to **us** vs. data **others** view as important.	The **sequence** by which we are exposed to data can change how we react to the **same** data.

It gets worse! Even from the **same set** of data, different people will, of course, see different possibilities and draw differing conclusions.

So, how can we deal with all this? Try **collective** wisdom!

Collective Wisdom

Most people, even those who've lived long enough to gain some wisdom, have **limited** perspectives, perceptions and resources. So wisdom is often more valuable if it is the **collective** wisdom of a group or team:

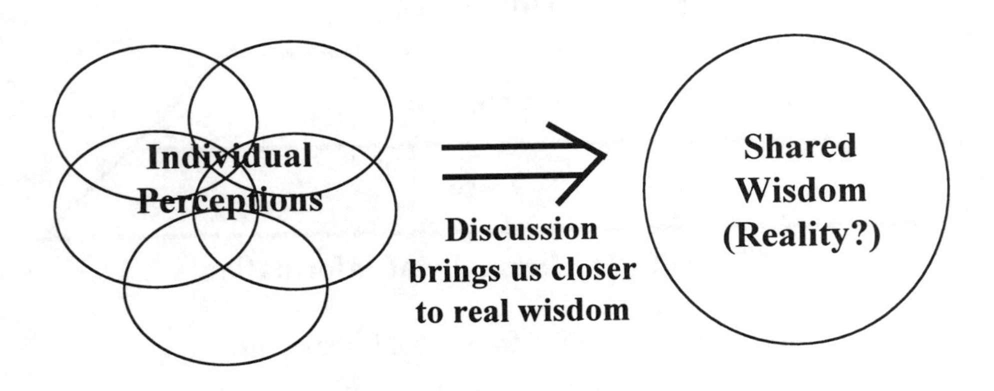

Figure 18.2: Collective Wisdom

Sharing and discussing perceptions helps us as individuals get around some of the data traps and faulty reasoning that interfere with individual wisdom.

For us as individuals, whether we're challenged by problems at work or at home, we can benefit from **collective wisdom!** This is truer and more critical now than at any time in history. Here are two examples to illustrate my point:

⇒ **A client executive who was CEO** of a large firm, is a very astute business person—well beyond the norm. Yet his Achilles' heel is that he tends **not** to "wander around" (as in MBWA). He welcomes inputs if offered, but doesn't seek out the perspectives of his staff or others. His tenure has been full of nasty business surprises, tough on him, but tougher still on his firm.

⇒ **Our office services consultant** (Ann Somboretz, Professional Office Services) on the other hand, has and uses a wide range of information sources on any topic of interest to her as well as her clients. Her sources include her clientele, her friendship network of business people, her family, and appropriate use of the Internet. Regarding the Net, I'm often surprised at the valuable information she turns up.

You may be fortunate enough to be part of several teams at work, to be part of a circle of friends or a support group in your "leisure time," and to be a member of a supportive family. If so, you have many opportunities to pursue **"collective wisdom"** in those settings. So, remember this sketch …

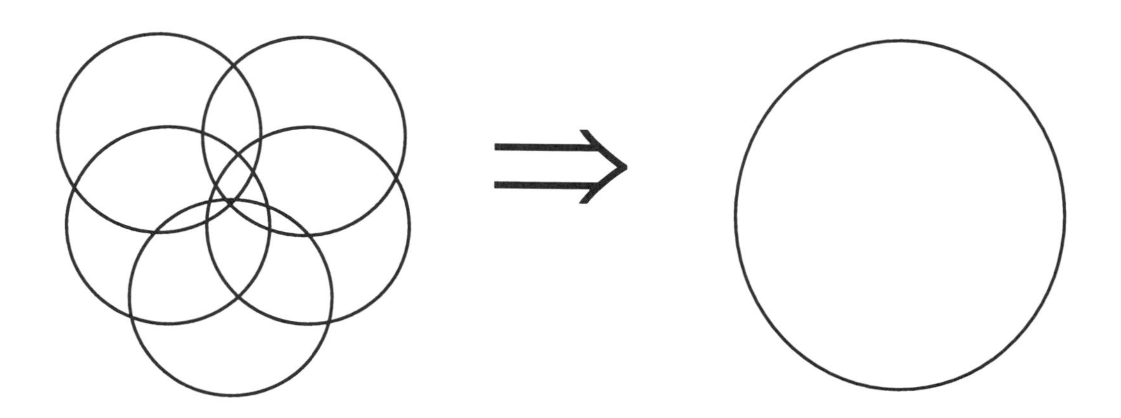

And remember the STOP Process described in Chapter 17, *Figure 17.1*.

Reflection

⇒ Identify several problems—work-related or personal—which would benefit from the insights of others.

Problem:	Sources of Wisdom:

Chapter 19

CRITICAL SKILLS OF TEAMWORK

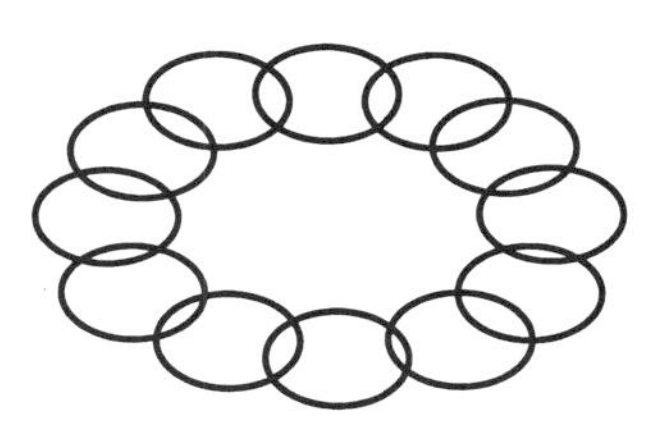

As the workplace changes, in every area of endeavor, there are very few things you can count on for the long term. But one of those few things will most certainly be the increasing need for **teamwork, interdependence and team-related skills**.

The Stakes for Teamwork

We've seen so much evidence that teamwork pays off in amateur and pro **group** sports, such as volleyball, soccer and hockey, as well as basketball, football, and baseball. Groups that have a high level of teamwork are inevitably more successful than those with less.

However, in a **work setting**, we often lose sight of the incredible value of teamwork, yet the stakes are far higher than in sports. It helps me to think of teamwork and the "stakes" involved in **workplace** teamwork using four "levels." They begin with sports and recreation and go from there.

Table 19.1: Workplace Teamwork "Stakes"

LEVEL	ACTIVITY INVOLVED	THE NATURE OF WHAT'S AT STAKE (THE "STAKES")
	Sports and Recreation **(Level One)**	Our team wins and loses, creating a common topic of **social conversation**.

Table 19.1: Workplace Teamwork "Stakes" *... continued ...*

LEVEL	ACTIVITY INVOLVED	THE NATURE OF WHAT'S AT STAKE (THE "STAKES")
	Everyday Interaction at Work **(Level Two)**	Our work team does well or poorly, impacting our **bottom line**, our compensation and perks, and possibly our **employment**.
	Health Care Team Members **(Level Three)**	Our diagnosis, surgical procedure, prescription, x-ray, therapy or whatever, goes well or poorly, impacting our **health**.
	Work at the Construction Site; Assembly, Power, or Process Plant **(Level Four)**	Depending on the skill and safety of our own work practices and those of our colleagues, our suppliers, etc., **our very lives are at stake.**

An example: Our son Terry, now in the military's special forces, regards his present employment as safer than his former work. Before the Gulf War, he was a carpenter foreman, working on pitched roofs, under construction cranes, and often working with people who were poorly trained and/or substance abusers!

Those who follow or play team sports are keenly aware of team skills at **level one**. Thanks to television dramas, or personal experience, we appreciate health care team skills (**level three**). Those who've worked on a construction site or factory/plant setting know first hand about **level four**.

Level two is the most misunderstood, neglected and under-appreciated. Yet, it has the highest potential payoff for team skills. We need to look at this potential through two sets of glasses:

- The group or **team**
- The group or team **member**.

The Nature of Teams

It is important to keep in mind that a team is a means to an end—an approach for achieving a goal, whether that goal is improved production, increased quality, better morale, happier customers, greater safety or worker satisfaction.

No matter what work is done, each effective work team has one thing in common: The need for "rules" to govern itself. Deborah Harrington-Mackin (***The Team Building Tool Kit***, AMACOM, NY, 1994) found that rules play a crucial role in **any** team's success.

Whether you call them guidelines, norms, ground rules, principles, standards or commitments, these team rules are critical, and ...

- rules are usually determined in the early months of a team's development, and, once established, they are difficult to change.
- any changes in team rules require substantial time and often cause team members to get upset.
- the team leader plays an important part in the setting of rules; what the leader doesn't do can be as important as what the leader does do.
- teams usually judge their members by how closely they conform to the rules, whether implicit or explicit.
- the more team members work together to develop their own team rules, the more they will agree with each other.
- a team willing to create rules is a team willing to be self-disciplined and to assume responsibility for its behavior.

Sample Team "Ground Rules"

A business organization of about 30 folks decided they wanted to improve their team's climate and effectiveness, so they spent several hours of intense work to develop these "team ground rules":

1. We speak candidly within the team.
2. We respect other members and their jobs as valuable.
3. We are honest about deadlines; we honor deadlines; we say "No" when appropriate.
4. We offer praise, encouragement, and positive reinforcement.
5. We are accepting of criticism; we are constructive when offering criticism.
6. We don't badmouth team members or others.
7. We strive for clear communications both in sending and receiving.
8. Regarding communication on "sensitive matters":
   - we are aware of our tone
   - we use e-mail carefully
   - we take it to those involved, not to a third party.

9. Regarding team goals and objectives:
   - we are explicit about them
   - we take personal responsibility for them.

After developing these excellent team ground rules, they reviewed them and committed to follow them as individuals. Not satisfied with "promises," they rated their performance on a scale of 0 to 5 over a 15-month period (see *Table 19.2*). Their self-ratings improved 1.6 on a 5 scale in that time-frame! And, more importantly, it's now a far more satisfying place for them **and** their clients.

Table 19.2: Self-Ratings on Team Ground Rules

(On a scale of 0-5, 5 being the highest)

	2/20/97	5/30/97	5/30/98
Rule #1	*1.8*	*3.2*	*3.5*
Rule #2	*1.9*	*3.2*	*3.7*
Rule #3	*1.7*	*2.8*	*3.1*
Rule #4	*2.6*	*3.5*	*4.0*
Rule #5	*2.1*	*3.3*	*3.8*
Rule #6	*1.6*	*3.1*	*3.5*
Rule #7	*2.3*	*3.4*	*3.5*
Rule #8	*1.8*	*3.0*	*3.6*
Rule #9	*1.8*	*2.9*	*3.6*
Overall	***2.0***	***3.1***	***3.6***

Stages of Team Development

My first book for ASCE, ***Collective Excellence: Building Effective Teams*** (1992) offered tips and tools for teams and team leaders. The one that seems to have proved most useful to many groups has been the **"Stages of Team Development."**

Here it is, reproduced with some slight editing from the 1992 version, as shown in *Table 19.3*.

Table 19.3: Stages of Team Development

Task groups, committees and work groups of all kinds may go through several stages of development. It pays to know what the stages are, what to expect and how to move on if you want to!

STAGE I COLLECTION (1)	STAGE II GROUP (2)	STAGE III DEVELOPING TEAM (3)	STAGE IV HIGH-PERFORMING TEAM (4)
People are ... • cautious • guarded • wondering Little visible disagreement The collection lacks an identity Little investment in the group function People are watching for the norms here to see what is okay or expected of them	**Group** is developing ... • identity • purpose • interest People are taking risks and getting to know one another Conflict is in nonproductive fits and starts High levels of frustration and/or confusion People develop pairs and cliques	**Emerging Team** is developing ... • goals • roles • relationships Members are learning to appreciate their differences Conflict is usually on issues, not about egos Communication is open and clear Sense of belonging Sense of progress Enjoying work	**The team** is acting on common goals with ... • synergy • high morale • high productivity Easy shifting of roles from one to another Differences are valued Looking out for one another's interests Spontaneous, collaborative efforts Sharing of all relevant information Conflict is frequent, often looks like problem solving

(0) (3) (6) (9) (12)

Improving Work Groups by Francis and Young provided the most insight into the development of this table, which is based on our consulting experiences.

Effective Team Members

Price Pritchett has written a useful, practical and compact handbook for team members that addresses a great need in the teamwork literature. He believes "The secret to teamwork lies in the team **members**," as he says in ***The Team Member Handbook for Teamwork*** (Pritchett Publishing Co., Dallas, TX, 1992).

That's true, he says, whether we're talking about basketball, surgery, fire fighting, music or drug busts. Here's a list of suggestions for people who want to be seen as **effective team members**, based on his work and ours.

- **Know your job** and do your best ("play your position").
- **Continue to develop** your work skills ("bring talent").
- **Support and use diversity** of talents and personalities.
- **Work hard at communication** (See Chapter 16).
- **Keep the big picture in mind**, so you can …
- **Help others** who need an assist from a teammate.
- **Bring real problems** to the team's attention.
- **Build up your teammates**; give recognition.
- **Watch your ego**; there's no "I" in teamwork.
- **Be a good sport**; use humor in positive ways.
- In public, **support team and leader decisions**.

Behaviors to Avoid

Some individual behaviors are a problem for the team and the leader and need to be avoided by each member. When they occur, they need to be brought to the offending member's attention.

There are probably thousands of **team-busting behaviors**, but here's a short list of frequent abuses to watch out for:

- Being stubborn to a point of frustration
- Being sarcastic, cynical and generally negative
- Covertly badmouthing others, creating rumors, etc.
- Engaging in horseplay, nonchalance, wisecracks

- Seeking personal recognition; "ego-flatulence"
- Being indifferent, aloof, non-participative
- Riding your personal agenda into the ground
- Manipulating the group in various ways.

Things To Ponder

As you think about an important group or team you serve on, what **dysfunctional** behaviors do you often see there?

____

____

____

____

____

____

____

____

____

For that same group or team, **what "team ground rules"** might improve productivity, communication, morale, quality of service or effectiveness?

____

____

____

____

____

____

____

____

____

How might **you** become a more effective team member or increase your contribution to teams you serve on?

Chapter 20

DEALING WELL WITH CONFLICTS

As the workplace continues to change, one thing we can **absolutely count on** is this:

MORE ISSUES! MORE CONFLICT!

So, it's becoming ever more important to have some functional perspectives and effective skills for **managing and resolving work place conflicts and personal life conflicts**.

Much new work has been done in conflict management in recent years, from "partnering" to "facilitation" to "win/win" approaches and more effective personal counseling. However, at the individual level, skills need to be improved even faster. And, they are becoming ever more necessary for career success.

Why Conflict Is so Challenging

For many people, disagreements, hassles and conflicts are a **relatively small percentage** of daily interactions. Why, then, are they so difficult, frustrating, painful and challenging?

Because they often arise from or cause intense emotions!

When I've assisted a management team or group in learning more about handling their conflicts, it's been helpful to ask them to honestly report the **emotions** they often feel about, during or after conflicts. No matter what kind of group I ask, they produce a list of emotions similar to this:

scared	annoyed	furious	wronged	outraged	put-down
worried	perplexed	sad	guilty	misunderstood	set-back
fearful	hurt	confused	stupid	shocked	sorry
angry	disappointed	humiliated	blocked	wasted	vengeful

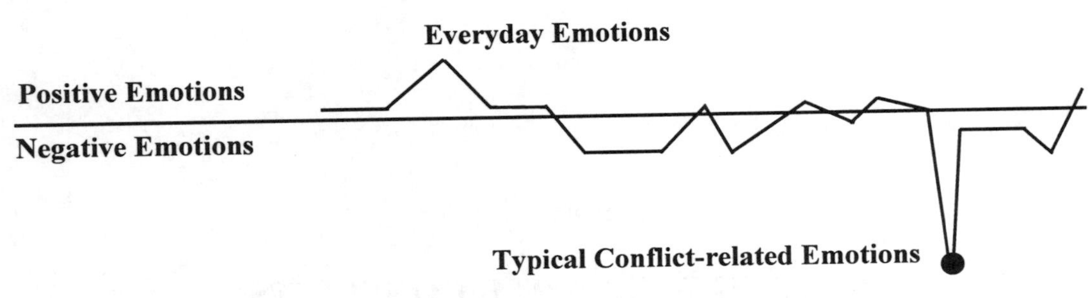

Figure 20.1: Emotions

> So, clearly we need some simple ways to be effective in conflicts in spite of strong and often negative emotions!

Diagnosing Conflicts

There are at least six different sources or "drivers" for conflicts. Generally, each different **source** of conflict will respond best to a **different strategy or approach**. *Table 20.1* provides some insight into categories of conflict and responses that may prove helpful. They are listed generally in order of increasing difficulty.

Table 20.1: Sources and Categories of Conflict

SOURCE OR CATEGORY	POSSIBLY HELPFUL STRATEGIES, APPROACHES OR RESPONSES
Different needs, interests or wants.	☆ Identify **each** party's interests, needs or wants. ☆ Avoid discussing "positions" (the "right way" to solve it). ☆ Brainstorm **lots** of ideas for creating a win/win solution.
Built-in ... ☆ role binds ☆ time constraints ☆ unequal resources, etc.	☆ Identify existing role perceptions. ☆ Clarify roles and responsibilities. ☆ Look for ways to **change** ... • perceptions • roles • inequities • structure, etc.

Table 20.1: Sources and Categories of Conflict ... *continued ...*

SOURCE OR CATEGORY	POSSIBLY HELPFUL STRATEGIES, APPROACHES OR RESPONSES
Information problems ... ☆ lack of data ☆ misinformation ☆ differing interpretations of data.	☆ Agree on how to gather needed data. ☆ Identify criteria for assessment. ☆ Analyze the data together. ☆ Clarify areas of disagreement. ☆ Use **outside** expertise.
Differing ... ☆ values ☆ beliefs ☆ principles ☆ cultures, etc.	☆ Identify common values, beliefs, etc. ☆ Look for common needs and interests. ☆ Create new areas of interest. ☆ Try to appreciate others' values. ☆ Agree to **respectfully** disagree.
Personality and relationship conflicts ... ☆ stereotypes ☆ past behaviors ☆ past emotions ☆ misperceptions ☆ different temperaments, etc.	☆ Create "ground rules" for discussion. ☆ Allow for venting pent-up feelings. ☆ Acknowledge feelings without judging. ☆ Surface unstated feelings or problems. ☆ Use an outside resource. ☆ **Only** when feelings are fully vented, begin problem-solving.

Differing Styles

Wallen, Blake, Mouton and others have identified five **basic styles or behaviors** that people seem to choose for handling conflicts. Briefly, these are:

☆ **"Sturdy Battlers"** push hard for their position and want fast resolution, preferably **their** way.

☆ **"Friendly Helpers"** put the relationship ahead of their needs and interests, often feeling "run over" as a result.

☆ **"Compromise-Seekers"** expect to bargain and reach a workable compromise through bartering or trading.

☆ **"Conflict-Avoiders"** see conflict as useless, punishing or a waste of time, and simply don't engage.

☆ **"Problem-Solvers"** use **patience and creativity** in trying to meet the needs and interests of **all** parties.

> **Caution:** Many of us see ourselves as "problem-solvers." However, when we ask our colleagues how **they** see our style, it's often something **quite different** than our view.

Really Helpful Tips

For business and family issues, the following may prove helpful:

☆ In most cases, if not all, it's often useful to solicit and take seriously what the **other party needs** or wants.

☆ Regarding **your own needs**, if your communication isn't working, **try something else** (use different approaches).

☆ If there are several issues, decide on one and **stick to it** rather than jump around. Don't complicate and obfuscate the discussion.

⇒ Most important of all, in most cases, **maintain rapport**. Rapport in a conflict situation includes:

- careful listening
- acknowledging the other
- respectful responses and comments
- matching eye contact, pace, voice tone and level
- avoiding personal put-downs or attacks.

☆ When emotions are running high, try a cooling off period.

☆ When these don't help, get appropriate outside assistance.

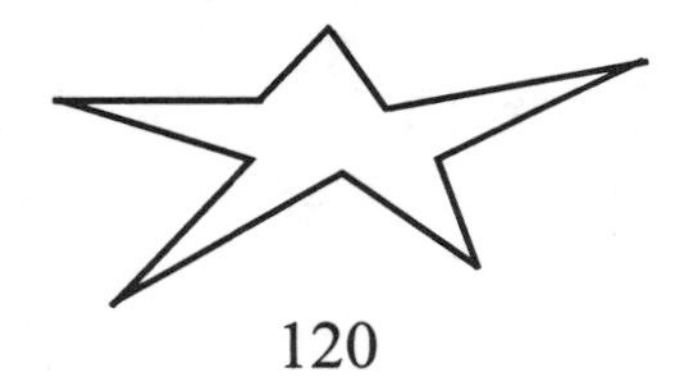

Predictable Multi-Group Dynamics

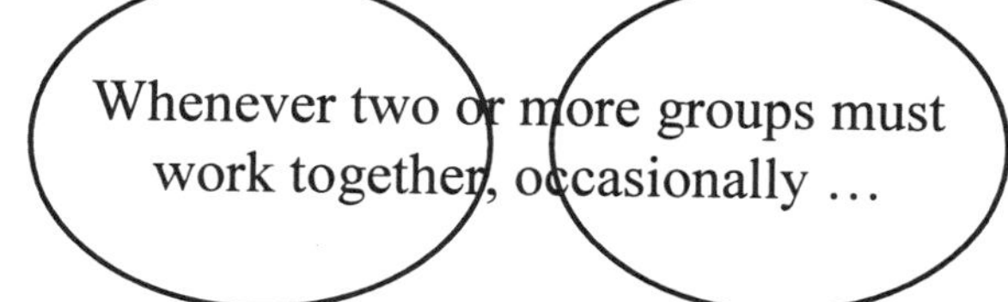

… they will experience almost predictable dynamics that are troublesome:

Typical Dynamics:

- ☆ The groups will almost always become competitive or **WE←→THEY**.
- ☆ Members notice **the worst** in other groups and minimize the positives.
- ☆ They may view their **own members as turncoats** if they comment positively about the "other" group(s).
- ☆ As time proceeds, it **amplifies itself**.

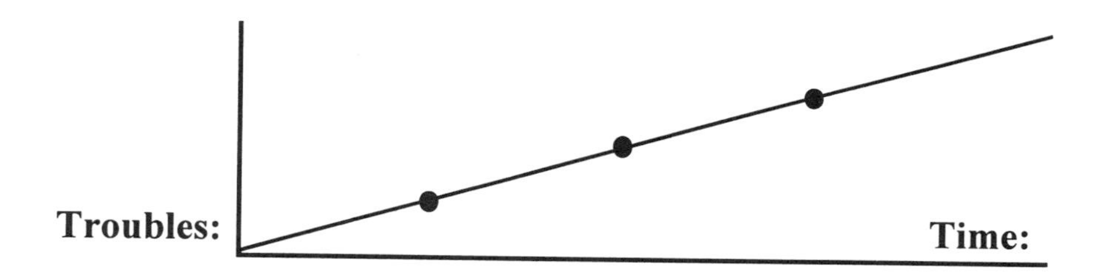

- ☆ If not interrupted, the groups will veer **farther and farther from reality** in their views of the other group(s)—to the point of **demonizing** the other.

Effective Interrupters:

- ☆ **A respected senior leader** who asks both groups to **work together more and better**.
- ☆ **"Turncoats"** who speak well of the "others" **to their own group**.
- ☆ **"Tourists"** who work in "other" group(s) for short or long periods.
- ☆ Opportunities to gather and talk to one another—both **socially** and on **real work issues**—with some frequency.

INDEX